A Non-Traditional Guide to Physical Chemistry

Robert Schiller

A Non-Traditional Guide to Physical Chemistry

Insights using Hydrogen

 Springer

Robert Schiller
Eötvös Loránd Research Network
Centre for Energy Research
Budapest, Hungary

ISBN 978-3-031-07490-5   ISBN 978-3-031-07488-2   (eBook)
https://doi.org/10.1007/978-3-031-07488-2

Translation from the Hungarian language edition: *Hidrogén, az elemek királya: A kémia születése, az ener-getika jövője* by Robert Schiller, © Typotex Publishing 2013. Published by Typotex Publishing. All Rights Reserved.
Language editor: Bálint Gárdos
This Springer imprint is published by the registered company Springer Nature Switzerland AG
The registered company address is: Gewerbestrasse 11, 6330 Cham, Switzerland

DMS

Preface

The book now in your hand is a most incomplete piece of work. And not only because it is much too short. Nowadays a chemist can claim to have understood a certain area of chemistry if some physical explanation of his observations has been found. Skill in mathematics, knowledge of physics, patience with definitions, statements, deductions, proofs, and falsifications are required for that sort of understanding. Chemistry is sometimes regarded as a branch of physics which deals with molecules or molecular assemblies. Physical chemistry has become an exact discipline. A good indicator of this development is that textbooks have become shorter: few basic ideas are sufficient to explain a number of effects.

That is not the path this book would follow, it is not my aim to discuss any problem in its entirety. Rather, I collect a series of examples in order to demonstrate the way how physical chemists think. The construction of the bridge which stretches between chemical experience and physical knowledge.

Hydrogen is the simplest of all substances. It is not simple to tell what simplicity means. It certainly does not mean that hydrogen is a cheap product which is easy to handle or that everything is made of it. It's much better to say that this is the substance whose behaviour is the easiest to explain through ideas and laws of physics. That is why I hope to be able to economize with words if physical chemistry is introduced by the example or pretext of hydrogen.

I shall use the discovery, properties, present-day applications and future benefits of this substance as a grain to be dropped into a beaker of science in the hope that some knowledge and understanding will crystallize on its surface. Thus I do not promise either strict rigour in exposition or thorough description in history. The perusal of the text is perhaps not too difficult because prior knowledge is usually not assumed and only little maths is used, the reader if she wishes so may even skip such passages. All these, of course, do not mean that the reader might be relieved of the burden of attention, concentration and understanding.

I believe that the essence of dealing with science is nothing else but wondering. I do so because of two separate reasons. One may be amazed if a well-known, apparently obvious phenomenon is in no connection with one's previous knowledge or understanding. And, on the contrary, one may be amazed that some unexpected experience can be understood in terms of well-known basic laws. Aristotle was certainly right when urged his disciples to wonder. The real aim of this book is to prompt the reader to wonder.

Budapest, Hungary   Robert Schiller

Contents

About the Author

Robert Schiller (1935), born in Budapest, Hungary, graduated from R. Eötvös University, Budapest, in 1958 and completed his Ph.D. in 1966 and D.Sc. in 1974. He is a titular professor at R. Eötvös University and Dr. habil Privatdozent at Budapest Technical University. After completing his studies, Professor Schiller joined the Chemistry Department of the Central Research Institute for Physics, Hungarian Academy of Sciences, where he is now a Research Professor Emeritus. Having worked at several laboratories abroad, he e.g. spent a full year at the Paterson Laboratories, Manchester, UK. His main research interests are in radiation chemistry, electrochemistry and the theory of transport processes. Currently, he is investigating the effects of fast ions on metals. He has taught courses on radiation chemistry and statistical mechanics at R. Eötvös University and has published several textbooks in these areas. Apart from his research papers, Professor Schiller has also written books and a number of essays popularizing science. He was awarded the Wigner Prize by the Hungarian Academy of Sciences in 2001 and was voted the popular science author for the year 2012, and asteroid no.196005 was named Robertschiller in his honour.

Toward Science

1   Farewell to Alchemy—Paracelsus

Some people think that hydrogen was discovered by *Paracelsus*. Perhaps this is not true. It is most probable, however, that no one would have discovered it for quite some time had he not been there.

Paracelsus, or Theophrastus Bombast von Hohenheim by his real name (Fig. 1), a scholar, living in the first part of the sixteenth century, an unexpectedly ebullient and fierce personality who never economized with words, rude as they might be, if the aim was to attack his fellow physicians or alchemists, but he contributed enormously to late medieval alchemy. "*Many maintain that alchemy is for producing gold and silver. Our present aim, however, is different: we wish to investigate the power and might of the medicines.*" Obviously, not all his predecessors believed that alchemy is nothing else but turning base metals into noble ones and he himself never denied that one can "produce gold". But his scientific interest, apart from his theological and mystical works, was centred on the preparation of compounds with certain healing effects. The sheer idea that a physician could use anything other than substances found in nature was quite a bold step, involving a breach with Galenic pharmacy, which used only herbs and plants for healing. The old method was rooted in late antiquity and was named after the Greek physician *Galenos*. Paracelsus used the products of his laboratory for healing the sick and often successfully, being wise enough to realize that it is the dose which makes difference between medicine and poison.

© The Author(s), under exclusive license to Springer Nature
Switzerland AG 2022
R. Schiller, *A Non-Traditional Guide to Physical Chemistry*,
https://doi.org/10.1007/978-3-031-07488-2_1

Fig. 1 Portrait of Paracelsus, after a work by Augustin Hirschvogel, from 1540

Even the fierceness of the scientific dispute about this innovation can demonstrate how radical this step was. Paracelsus' opponents were similar to him if he was about to hurl abuses stating that the new medicines ravage more dangerously than the pest itself. (Perhaps the antimony preparations were too generously dispensed.) This left a mark also on the drama. Goethe's *Faust* could have been a contemporary of Paracelsus. Faust connects the alchemist's manipulations to his father, who also tried to produce medicines and concludes "*So we roamed, with our hellish pills, / Among the valleys and the hills, / Worse than the pestilence itself we were.*" (translated by A. S. Kline). Several further ideas of Paracelsus, judged most severely in these lines, appear in the drama.

Indeed, Paracelsus' fame is due much less to his medicines than to his conviction that medicines are not to be found but prepared. Research must show which preparation can heal a given illness. The rational order of the researcher's thinking is still unknown to him. The researcher's motives and disposition, however, appear already in his works.

The present-day reader might easily be deterred from his works. Luckily enough the original Old and Medium High German texts interwoven with Latin semi-sentences are nowadays available in a modernized version. Obviously, no modernization of the language can ease the understanding of the complicated descriptions of the alchemists' manipulations or of the hazy mysticism of their theories. Many of his present-day readers, people with only limited scientific interest, hail the great mystic in his person. It is, however, probably more than mere chance that his most powerful notion, the idea of the *tria prima* (the three primordial principles), which had a long lasting effect through the mysticism of *Jakob Böhme* and *Angelus Silesius*, was based on sober observations made in the laboratory.

The first thing an alchemist observed was the state of the substance. The present-day notion of chemical composition being unknown to them, their analysis mostly consisted in destructive distillation. That is the reason why the alchemist's workshop was full of furnaces, receivers, alembics, cauldrons, and stills (Fig. 2). In these early days substances were characterized by their most obvious property, their state. What we must regard as naïve in this characterization is the belief that in the course of chemical manipulations states are conserved in a similar sense as we know nowadays charge or mass to be conserved.

The three primordial principles correspond to the three states of matter. "*...whichever body consists of three things. Their names are: sulphur, mercurius, sal ... Taking a body in your hand you hold three substances in a single form.*" Any Latin dictionary can tell us that 'mercurius' obviously means mercury and 'sal' means salt. But stop smiling at this system of analytical chemistry and continue reading the First Book of *Opus Paramirum*. "*Whatever burns is sulphur, nothing else burns but sulphur; whatever smokes* (evaporates) *that is mercurius, nothing else does sublimate but mercurius. Whatever turns into ash is sal, nothing else turns into ash but sal.*" The effects observed during heating and distillation are summarized by these three names. Paracelsus makes clear that he does not mean actual substances stored on the shelf. "*Mercurius is the spirit (spiritus), sulphur is the soul (anima), sal is the body (corpus).*" The alchemist's art, called *scientia separationis* (the science of separation), recovers these principles from tangible bodies. Moreover, accentuating that he means effects rather than bodies, he explains a few lines below that the yellow crystals in the crucible are *similar* to sulphur, mercury to mercurius, salt to sal, because their effects are similar. Their effects? One would rather say their behaviour. The principle of inflammability is best represented by sulphur,

Fig. 2 Woodcut from Lazarus Ercker "Beschreibung allerfurnemisten mineralischen Ertzt und Bergwercksarten" Frankfurt 1598 (after J.R. Partington, *A History of Chemistry*, MacMIllan, London 1961, by the permission of the Publisher)

volatility by mercurius, whereas the non-inflammable sluggishness appears most obviously in salt.

Substances, "bodies" are different only because they contain the three primordial principles in different proportions. Principles and not elements, neither in the Aristotelian nor in the contemporary sense of the word. "*There is a different sulphur in gold, again different in silver, different in lead, in tin … different in sapphire, in ruby, in chrysolite, in amethyst, in magnets…*" The list goes on and on—Paracelsus as an author is loquacious and undisciplined.

Having such a picture of material composition in mind, it is only too natural that he believes in the transmutation of metals as everybody else did in his age. According to their idea base metals slowly ripen to become noble in the Earth's womb. So the alchemist's workshop might also become appropriate for the same process to take place. Naturally, we know only too well that it is impossible, or … what is it exactly that we do know?

We know that some substances, if they are stirred, heated on a stove, or exposed to sunshine can be transformed into each other, whereas others cannot. Those substances which cannot be made lighter than they are we call elements. Experience over centuries has shown that no stirring, heating, or illumination can produce element from element. That is one of the basic tenets of post-alchemic chemistry.

Since *Hermes Trismegistos*, Paracelsus or even *Newton* (who also tried his hands at such a task) failed to make gold from lead or lead from gold, we can state that these metals are elements. By today it is clear that nuclear reactions can take place at the expense of much higher energies; that is, elements can be transformed. Nevertheless, we regard both gold and lead as elements.

People were desirous of making gold because it was the common medium of exchange due to its appropriate physical properties, nice colour and relative rarity. Anyone who had a lot of gold was a rich person. We look down on the alchemists because they had only low energies at hand and the common medium of exchange happened to be a chemical element. But at certain times in China a gold coloured mineral, called cinnabarite (mercury sulphide), had a value higher than gold. Mercury sulphide can be produced in the alchemist's workshop, for example from mercury and sulphur. (As a matter of fact, mercury was produced from cinnabarite but let this fact be forgotten for a moment.) Had cinnabarite become the common medium of exchange the alchemist mirage would have turned into reality. Clearly, but this is an economical problem, a substance which can be synthetized without any limit could not serve as a medium of exchange.

Having related uncountable operations performed in his laboratory Paracelsus also describes the way he dissolved iron filings in oil of vitriol, i.e. in concentrated sulphuric acid (Fig. 3). Any present-day chemist would think it to be difficult to reproduce the experiment, among other steps keeping the alembic in manure (that was the thermostat of that age), nevertheless it is clear that the iron became dissolved. Although we all know that hydrogen evolves in the course of this process the text does not hint at any such event. However, a few pages earlier one finds the following sentence (Fig. 4). "*Thus keep it well in mind: the elements occur in the very same form and shape during the course of separation as they were present in the basic elements. Because air turns out to be similar to air, and indeed it cannot be gripped by hand, although some people think it in their minds that way. And the reason of that is that the air elevates in the instrument of separation, and breaks forth like the wind and sometimes it elevates with water, sometimes with earth, and again sometimes with fire. Because there is a wondrous elevation in the air.*"

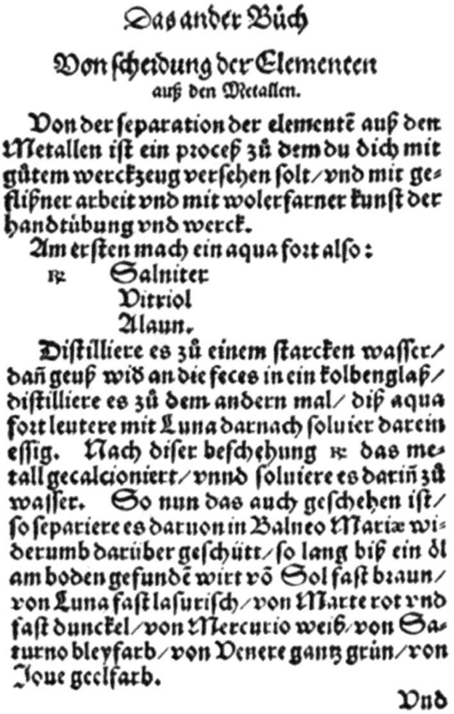

Das ander Bůch

Von scheidung der Elementen

auß den Metallen.

Von der separation der elementē auß den Metallen ist ein proceß zů dem du dich mit gůtem werckzeug versehen solt/vnd mit geflißner arbeit vnd mit wolerfarner kunst der handtübung vnd werck.

Am ersten mach ein aqua fort also:

℞ Salniter
Vitriol
Alaun.

Distilliere es zů einem starcken wasser/ dañ geuß wid an die feces in ein kolbenglaß/ distilliere es zů dem andern mal/ diß aqua fort leutere mit Luna darnach soluier darein essig. Nach diser beschehung ℞ das metall gecalcioniert/vnnd soluiere es dariñ zů wasser. So nun das auch geschehen ist/ so separiere es daruon in Balneo Mariæ widerumb darüber geschütt/so lang biß ein öl am boden gefundē wirt vō Sol fast braun/ von Luna fast lasurisch/von Marte rot vnd fast dunckel/von Mercurio weiß/von Saturno bleyfarb/von Venere gantz grün/von Joue geelfarb.

Vnd

Fig. 3 Page from the book *Archidoxa* by Paracelsus with the description of iron dissolution in sulphuric acid (Library of the Hungarian Academy of Sciences)

Our minds would find it difficult to think that this description is precise. The statement about the air which is similar to air resembles some magic ditty more than it does a scientific statement. Nevertheless, the text testifies to a good number of observations on the formation of gases and vapours. All these in the close neighbourhood of the reaction between iron and sulphuric acid. That is the reason why several historians of science are of the opinion that Paracelsus must have observed the formation of hydrogen during some of his experiments, notwithstanding the lack of the explicit description of the process.

Whether or not that is true, this lynx eyed observer of material states was greatly entertained by the properties of "air", that is of gases and vapours. Just to mention a few of his observation, he realized that wood cannot burn if air is not present, also that animals suffocate without air. He was also the first to use the general name "*chaos*" for this kind of substances. From that, chemist *van Nostrand* in the Netherlands formed the word *gas* at the beginning of the seventeenth century.

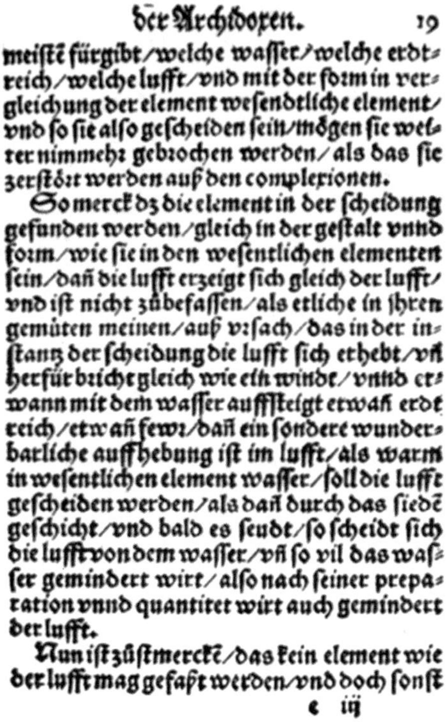

der Archidoxen. 19

meiste fürgibt/welche wasser/welche erdt=
reich/welche lufft/vnd mit der form in ver=
gleichung der element wesendtliche element/
vnd so sie also gescheiden sein/mögen sie wei=
ter nimmehr gebrochen werden/als das sie
zerstört werden auß den complexionen.

So merck dz die element in der scheidung
gefunden werden/gleich in der gestalt vnnd
form/wie sie in den wesentlichen elementen
sein/dañ die lufft erzeigt sich gleich der lufft/
vnd ist nicht zübefassen/als etliche in ihren
gemüten meinen/auß vrsach/das in der in=
stantz der scheidung die lufft sich erhebt/vñ
herfür bricht gleich wie ein windt/vnnd er=
wann mit dem wasser auffsteigt etwañ erdt=
reich/etwañ fewr/dañ ein sondere wunder=
barliche auffhebung ist im lufft/als wann
in wesentlichen element wasser/soll die lufft
gescheiden werden/ala dañ durch das siedẽ
geschicht/vnd bald es seudt/so scheidt sich
die lufft von dem wasser/vñ so vil das was=
ser gemindert wirt/also nach seiner prepa=
ration vnnd quantitet wirt auch gemindert
der lufft.

Nun ist zü stmerckẽ/das kein element wie
der lufft mag gefaßt werden/vnd doch sonst
c iij

Fig. 4 Page from the book *Archidoxa* by Paracelsus with the observation of gas evolution (Library of the Hungarian Academy of Sciences)

Given our present-day atomic-molecular model, the opposite of the strict atomic order is represented in a crystal by the structure of gases. Their atoms or molecules move completely at random, obeying nothing else but the laws of statistics. Did this forefather of us chemists, with his great intuition, suspect anything like that? Or did he just want to express the fact that these substances cannot be gripped by hand or poured in a vessel but they fly off as they wish?

The elements which man is made of are non-destructible. All that is received from Earth will return to Earth and there it remains until Sky and Earth come to an end; and anything that has water in him turns into water again, and this can be halted by no one; and that has chaos in him returns into the air, and his fire into the heat of Sun.

2   Fruitful Doubts—Boyle

To the best of our knowledge *Turquet de Mayerne* was the first observer who described the formation of hydrogen in the first part of the seventeenth century. I quote his own description from the English translation of the Latin text by J.R. Partington: "*I have taken 8 oz. of iron filings and in a deep glass cup I have added successively 8 oz. of oil of vitriol and a little later an equal quantity of warm water. There was produced an enormous agitation and a great ebullition and a meteorism of matter easily quieted by stirring by a rod. There is also raised a most fetid sulphurous vapour very noxious to the brain, which (as happened to me not without danger) if brought near a candle takes fire…*". Apparently, the gas was contaminated; pure hydrogen is completely odourless.

Chance and the almost merry playfulness of experimenting shine through the above description. Not much time after Turquet, but probably independent of him, *Boyle* performed the same experiment. If nothing else but this description would be extant of all of his achievements this alone would convince his reader that he started a new age of chemistry where the phenomena are investigated with resolution in their own right. He did not let loose the evolved gas but collected it over water, estimated its thermal expansion, finding it similar to that of air. And what is perhaps the most important point, he repeated the experiment under reasonably varied conditions. That is how a modern chemist works.

That experiment is just a tiny part of Boyle's vast oeuvre. Conducting a secluded, deliberately modest way of life, as the only member of an aristocratic family without any rank or title, he spent his whole time experimenting and penning or dictating the results. He was not very frugal with words and the malicious posterity maintains that his complete works have been read by his printer alone.

Besides his much-quoted gas law he is best remembered for his book entitled *Sceptical Chymist*. The chemist who, as stressed by the title of his book, being a sceptic, is decidedly antipathetic to the enigmatic wording and, even more so, the loose concepts of Paracelsus and his followers. "*… their writings, as their furnaces, afford as much smoke as light*".

The word sceptic expresses a firm philosophical stance. The classical Greek philosophical school of scepticism was founded by *Pyrrhon* who never put down his ideas in a written form. *Sextus Empiricus* was one of the authors who expounded the concept in great detail around the end of the 1st or the beginning of the second century A.D. At the outset of his work he made clear the aim of the school. "*Scepticism is an ability, or mental attitude, which opposes appearances to judgments in any way whatsoever, with the result that, owing to the equipollence of the objects and reasons thus opposed we are brought firstly to a state of mental suspense and next to a state of "unperturbedness" or quietude."* The sceptics maintain that all these can be attained if we refrain from making any judgement whatsoever. This is a wise thing to do because our personal experiences are basically unreliable and our statements are influenced by traditions and prejudices in our thinking. Thus, any judgement is built upon shaky foundations.

Boyle, the experimenter who is conscientious in putting down his experiences, is wise enough to doubt the strength of sheer observations. *"I look upon the common operations and practices of chymists, almost as I do on the letters of the alphabet, without whose knowledge it is very hard for a man to become a philosopher, and yet that knowledge is very far from sufficient to make him one."* Particularly he feels the meaning of the three primordial principles, mercury, sulphur and salt to be most vague. *"And indeed I fear that the chief reason, why chymists have written so obscurely of their three principles, may be, that not having clear and distinct notions of them themselves, they cannot write otherwise than confusedly of what they but confusedly apprehend."* All these, however, must have an end nowadays. *"I observe that of late chymistry begins, as indeed it deserves, to be cultivated by learned men, who before despised it."*

The definition of the element, proposed by Boyle, is indeed very clear and is closely related to our present way of thinking. *"I mean by Elements, as those Chymists that speak plainest do by their Principles, certain Primitive and Simple, or perfectly unmingled bodies; which not being made of any other bodies, or of one another, are the Ingredients of which all those call'd perfectly mixt bodies are immediately compounded, and into which they are ultimately resolved."*

A "perfectly mixt body" is, of course, a compound, in contrast to some imperfectly mixed, mechanical mixture. The above definition of an element is otherwise very old being akin to the formulation of ancient stoic philosophy—at least as far as wording goes. This definition mostly meant a dictionary-like interpretation or paraphrase of the word for the stoics. Knowing, however, the wealth of experiences at Boyle's disposal the definition for him must have served as an organizing idea of innumerable observations and deliberately generated chemical transformations. Unfortunately, the continuation of the text is somewhat equivocal; thus, the reader is unable to find out which of the substances were taken as elements by him.

Boyle endorsed atomic theory as he understood it from *Gassendi*'s book about *Epicurus*; moreover, he maintained as a basic principle that the permanent motion of the particles is their inherent property. According to his idea, the primary particles, which are the smallest ones as well, gather to form small bodies, which behave like elements. At some points he uses the words corpuscle or small particle instead of atom. Probably not by mere chance. Rather, as he makes it clear, he is not going to take sides in the debate on whether matter is or is not infinitely divisible. Here he admits his ignorance. He found the notion of corpuscle or atom to be indispensable in the creation of mechanical philosophy. While he might have had doubts regarding the corpuscle's indivisibility he was sure of their existence and uninterrupted motion. He was no atomist in the strict sense of the word. He based his explanation of natural phenomena on the laws of mechanics and not on indivisible atoms. In antiquity it was thought that nature and machines work differently. Boyle was of the opinion that the corpuscles in a substance work as machines do: similar to levers, balances or pendulum clocks. He did not believe, however, that his corpuscular philosophy would offer a complete explanation of the world as a whole. Probably his religion also prevented him from that type of materialism.

As Newton's older contemporary it is only too natural that Boyle tried to explain his experiences in mechanical terms. Curiously enough he was, at the same time, suspicious of the increasing hegemony of mathematics. He regarded mathematics as the most abstract discipline, a science with no direct contact with the phenomena of nature. Did he not know or did he even deny Galileo's dictum, much quoted even nowadays, about the book of nature which is written in the language of mathematics?

In any case, he always tried to quantify his observations. The volume increase of water during freezing was observed only qualitatively: he found that the ice made thick walled metal vessels rupture. But the densities of substances were measured meticulously by making use of a hydrostatic balance or a "specific gravity bottle", similar to our present-day picnometers, and the results were given with a precision of three or four decimals.

He was deeply interested in the behaviour of air and gases; Paracelsus's ideas about the impossibility of gases to be gathered or manipulated had become obsolete by that time. He must have known the works of *Torricelli* and *Pascal* on atmospheric pressure and vacuum, similarly the air pump of *Guericke* together with the related experiments. He himself built a pump for the demonstration that liquids boil at lower temperatures when the air is thinner above their surfaces. He made use of this observation in the construction of a vacuum still.

His most important experiment, which makes his name a household word among physicists and chemists even nowadays, was performed with very simple means. A U-shaped tube closed air tight at one end was filled with mercury; thus, a certain fixed amount of gas was enclosed in the closed limb. The gas volume and pressure (i.e. the height difference of mercury in the two limbs of the tube) was measured. Performing the experiment with pressures both higher and lower than that of the atmosphere Boyle could state that his results agree closely "*with the hypothesis, that supposes the pressures and expansions* [volumes] *to be in reciprocal proportion*".

The formulation is exemplary. It refers to the inevitable experimental errors which preclude any observation to agree perfectly with a mathematical relationship, whereas this latter must be considered to be a hypothesis until it is verified through measurements. That is how a modern scientist is supposed to think. The experiment is also excellent, the Boyle-Mariotte Law (*Mariotte* re-discovered the relationship a few years later) is demonstrated at school in the same way. Few experiments survive more than three hundred years!

Denoting pressure by p and volume by V, using our modern formulaic language, the law states:

$$pV = \text{constant} \qquad (2.1)$$

One must keep in mind that the constant depends on temperature, a fact which was realized already by Boyle in view of his own experiments on heat expansion and also knowing Galileo's thermoscope.

Boyle was much revered already during his lifetime. Perhaps the highest of praises he obtained came from *Oldenburg*, the first secretary of the Royal Society, in a letter to *Spinoza* referring to a scientific dispute: "*Our Boyle is not of the number of those who hold so fast to their own opinions that they do not need to take account of the agreement between them and the phenomena.*"

3   The Birth of Physical Chemistry—Lavoisier

Burning is the most spectacular chemical process. Throughout the history of the human race there has always been a story, told or written, about fire snatched from Heaven, a shower of fire, a sacred fire guarded by priestesses, a magic fire, or fire just as a basic element. The natural philosopher's most obvious task was to understand this very well-known, yet always amazing phenomenon. By the late seventeenth and early eighteenth century chemists amassed such an amount of experimental facts that they could try to formulate some unified theory of burning.

A unified theory! Having had ideas about heat, flames and ignition since times immemorial, the chemists' difficult task was to find a single principle by which one could account for that plethora of phenomena. It was easy for Aristotle to regard fire as a single element. Being an experimenter, Paracelsus saw different burning agents (different forms of the principle sulphur) in each and any inflammable substance.

This "multitude of sulphurs" was united into a single fire-like element by *Georg Ernst Stahl* based on the idea of *Johann Joachim Becher*. Stahl thought that there exists a substance, he called it phlogiston, which is a constituent of each inflammable, burnable substance the difference consisting only in the amount of that substance. (The expression, already coined before Stahl, had been derived from the Greek word *phlox*, meaning flame.) Combustion is nothing else but the exit of phlogiston from the inflammable substance. The "calcination" of metals by which "earths" are formed (in modern terms the oxidation of metals resulting in oxides) is a process similar to burning: phlogiston leaves the metal and the residue is the "earth" which is not inflammable anymore because it is barren of phlogiston. If a substance which does not contain phlogiston, i.e. a product of burning, is heated with another one, which is rich in phlogiston (like fat or charcoal) than phlogiston enters the product of burning, thus recovering the inflammable substance. Stahl burnt phosphorous and obtained phosphoric acid, out of which, by heating it with charcoal, he recovered phosphorous. Metals can be produced from their respective "earths" by making use of charcoal—this method has been well known since ancient times when the first metal furnaces were built.

Curiously enough, Stahl was aware of two important facts that would later be used as important arguments against his theory. He knew that burning needs air and he also knew that metals are lighter than their combustion products. As to the first fact, he stated that air is necessary for the phlogiston to escape but it does not participate in the process. As to the second one, the weight increase was of little concern to him because, not having ever seen pure phlogiston, he might have found it easy to think that its presence decreases the substances' weight. Moreover, in his time quantitative data were of minor importance in comparison to qualitative insights.

Nowadays, people are likely to smile upon the theory and looking down on its superseded naivety. Nevertheless, it made a great impact in its own time. The idea, which was easy to comprehend, offered a unifying view to all important experiences regarding combustion. Compared to Stahl, those contemporaries who rejected his theory were lagging behind in understanding the essence of these processes. *Boerhave*, the author of the excellent, descriptive work entitled *Elementa Chemiae* did not accept the phlogiston idea because he did not concede that burning and metal calcination are related processes. He accepted that there exists a combustible material, which leaves the sulphur as it is burning. However, he denied the existence of any important difference between a metal and its earth. It is held that this anti-phlogistic stance was the reason why the book by Boerhave, a most influential physician and university professor, was soon forgotten.

Phlogiston has not been forgotten. Few elementary text books can resist the supposed triumph of disproving that theory. Nevertheless, apart from the role it played in the development of chemical thinking, "there is undeniably something in it". At least the venerable age of that false idea is an indicator. Nowadays, we all learn that combustion is nothing else but composition with oxygen. Also, we learn that combustion is a specific case of oxidation, a process which consists in losing an electron by the substance being oxidized. For example, the burning of sulphur is described by the following chemical reaction equation,

$$S + O_2 \rightarrow SO_2 \tag{3.1}$$

and the oxidation of mercury by this,

$$2Hg + O_2 \rightarrow 2HgO \tag{3.2}$$

As sulphur or mercury undergoes this reaction they give a negative charge, that is an electron, to the oxygen. The process writes as,

$$2Hg + O_2 \rightarrow 2Hg^{2+} + 2O^{2-} \tag{3.3}$$

If the oxide is made to react with carbon, hydrogen or any other reducing agent in order to recover the metal, the essence of the process is the transfer of an electron to the oxidized substance. We write this as

$$Hg^{2+} + O^{2-} + H_2 \rightarrow Hg + 2H^+ + O^{2-} \tag{3.4}$$

Finally, the positive hydrogen ions and the negative oxygen ion form water,

$$2H^+ + O^{2-} \rightarrow H_2O \tag{3.5}$$

(Electrically charged atoms or molecular particles are called ions.)

Nowadays one may write that "combustion is nothing else but an electron leaving the burning substance" or that "if the combustion product is heated with a substance of high electron density electrons go over to the product and the combustible gets recovered". A present-day chemist would accept these statements perhaps just finding the wording somewhat weird. I do not suggest that the twentieth-century theory of oxidation–reduction processes was presaged three hundred years before. Even less do I propose phlogiston to have been an early model of valence electrons.

Phlogiston theory was obviously erroneous. But it was a deep furrowing error.

The earlier part of the eighteenth century, the time before *Lavoisier's* activity, is often called the age of pneumatic chemistry, the chemistry of gases. The key word is chemistry because scientists like Paracelsus, van Helmont or Boyle, to name but the greatest personalities, made a good number of observations on the behaviour of gases. They, however, marvelled at the physical properties, hardly realizing that the gases produced by different methods also differ in their chemical character. It came only with the ingenious experiments of *Black, Cavendish, Daniel Rutherford, Scheele* and *Priestley* that one could realize, for example, that air is a mixture of two kinds of gases and find a difference between the components by observing that only one of them feeds combustion; or that one could see that the gas which is set free in a lime kiln is different from the one which evolves when iron filings meet sulphuric acid. Indeed, with such observations at hand there was a brief period when pure phlogiston was thought to have been produced; however, the results of these very experiments were compelling enough to discard the idea of phlogiston.

Magnesium alba i.e. magnesium carbonate was heated by Black, but he was not the first chemist who tried this simple experiment. Nevertheless, he recognized that the nature of the evolved gas differs from that of natural air: the two act differently on lime water (aqueous solution of calcium hydroxide). Also he realized that the gas which comes from magnesia alba is also a minor component of natural air. Called "fixed air" by Black that gas is known to us as carbon dioxide.

Let us make use of present-day chemical language. When magnesium carbonate is heated, carbon dioxide is set free,

$$MgCO_3 \xrightarrow{\text{heat}} MgO + CO_2 \qquad (3.6)$$

which is absorbed by lime water,

$$CO_2 + Ca(OH)_2 \rightarrow CaCO_3 + H_2O \qquad (3.7)$$

Even apart from its small carbon dioxide content natural air was shown to be a composite substance. Daniel Rutherford burnt phosphorous in a closed vessel, absorbed the smoke in lime water and still he found a certain amount of gas in the vessel. He called this component "noxious gas" or "phlogisticated air", a logical name within the framework of Stahl's idea: air can feed combustion until it can offer some place for phlogiston. As the air gets saturated with phlogiston burning stops.

Nowadays the burning of phosphorous writes as

$$4P + 5O_2 \rightarrow 2P_2O_5 \qquad (3.8)$$

The smoke consists of phosphorous pentoxide which reacts with lime water,

$$P_2O_5 + 3Ca(OH)_2 \rightarrow Ca_3(PO_4)_2 + 3H_2O \qquad (3.9)$$

Both D. Rutherford and Scheele, who made similar experiments, found it difficult to saturate air with phlogiston. A mouse suffocated in the vessel, so the atmosphere was deemed to be "full of phlogiston", but a candle burnt after a while and even then "there remained some room for phlogiston to be obtained from phosphorous".

Scheele found the fraction of air which feeds burning to be one fourth of the total volume and called it "fire air"; its later name was "dephlogisticated air" since it fed combustion, which was seen as a clear sign of its ability to absorb phlogiston.

Scheele prepared "dephlogisticated air" in pure form by heating the mineral pyrolusite with sulphuric acid. We write this reaction as

$$MnO_2 + H_2SO_4 \rightarrow MnSO_4 + H_2O + 1/2O_2 \qquad (3.10)$$

Priestley's famous experiment consisted in heating "calx of mercury" (red oxide of mercury) by focusing sunlight and finding the formation of "dephlogisticated air". A present-day chemist understands this process as the thermal decomposition of the oxide, writing

$$2HgO \xrightarrow{\text{heat}} 2Hg + O_2 \qquad (3.11)$$

Thus, oxygen was discovered by Scheele and Priestley.

If "phlogisticated air" and "dephlogisticated air" were mixed in appropriate proportion, the density and chemical behaviour of the mixture was found identical to those of natural air.

Cavendish, called "the richest of all the savants and the most knowledgeable of the rich", being the archetype of the secluded English aristocrat, was the first to produce pure nitrogen and describe it with precision. This achievement is but a minute part of his great oeuvre in chemistry and physics. Having made natural air pass over red hot charcoal several times he washed the gas with some alkali in order to remove carbon dioxide. He also measured the gas density by weighing an animal bladder filled with the gas. He found that the component of the air which does not feed combustion is somewhat lighter than air.

These experiments, and a good many others, lent a great momentum to the science of chemistry not only by their results but also by the methods employed. An important improvement was, for example, Priestley's idea to collect gases in a vessel closed by mercury instead of water in order to also investigate water soluble gases.

Also "inflammable air", the gas we now call hydrogen, was studied by Cavendish following the experiments of Turquet de Mayerne and Boyle. This research put in order all the previous results as far as both preparative methods and quantitative relationships are concerned. Cavendish showed that, considering all the metals known by him, only iron and zinc can develop "inflammable air" under the effect of dilute "oil of vitriol" or "spirit of salt" (sulphuric acid or hydrochloric acid). We write the reaction simply

$$Zn + 2HCl \rightarrow ZnCl_2 + H_2 \tag{3.12}$$

He also recorded the weight of the metal and the volume of the gas not forgetting to note the ambient temperature and pressure. He found "inflammable air" to be lighter than normal air by a factor of 107/8 (the correct value being 14.4).

Writing about its composition, he stated that "[the metals'] phlogiston flies off, without having its nature changed by the acid, and forms the inflammable air". Both he and Priestley, his contemporary, were certain that the gas which evolves in the course of the zinc plus acid reaction is nothing else but free phlogiston. Sure, because the two products of the reaction are an incombustible salt and a gas which can burn fiercely. This was regarded as a great feat of preparative chemistry: Stahl's abstract looking idea became a substance which can be put in a vessel and has a definite volume, mass, and chemical properties.

The substance they denoted as "inflammable air" and thought to be pure phlogiston is called nowadays hydrogen.

"Inflammable air" which was thought to be pure phlogiston and "dephlogisticated air" can react with each other as it was found by Cavendish and a number of other chemists. This reaction is well known to us and writes as (Fig. 5)

$$2H_2 + O_2 \rightarrow 2H_2O \tag{3.13}$$

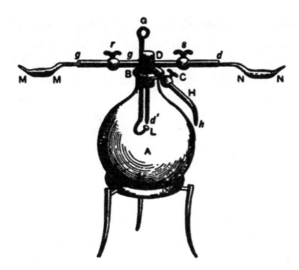

Fig. 5 Lavoisier's equipment for water synthesis form a mixture of hydrogen and oxygen by an electric spark (From J.R. Partington, A History of Chemistry, London MacMillan 1961, by the permission of the Publisher)

Cavendish having repeated the reaction and also supporting it with meticulous mass determinations arrived at the reaction scheme

$$\text{phlogiston} + \text{dephlogisticated air} \rightarrow \text{water} \tag{3.14}$$

Hence "dephlogisticated air" is nothing else than water deprived of its phlogiston.

This reaction equation makes the present-day reader bemused as it has been recognized by a number of historians. The explanation must lie in the change of the meaning of a reaction equation since Cavendish's time until now. He must have regarded water as an element, as Aristotle did, whereas "inflammable air" was thought rather a mixture of water and phlogiston than pure phlogiston. Perhaps the real meaning of the above equation is

$$\underbrace{(\text{water} + \text{phlogiston})}_{\text{"inflammable air"}} + \underbrace{(\text{water} - \text{phlogiston})}_{\text{"dephlogisticated air"}} \rightarrow 2\,\text{water} \tag{3.15}$$

One must suppose some similar idea to have prevailed, otherwise one would be compelled to believe that this brilliant experimentalist was unable to distinguish between air and water.

We find difficult to follow the way of thinking in that age. Despite the compelling evidence against phlogiston theory at hand it was still not thought to be absurd to identify phlogiston with the gas we now call hydrogen.

The theory was refuted by *Lavoisier* due to much deeper considerations. He was the first of the chemists who had a lasting co-operation with a physicist. He was lucky enough to have the great theoretician *Laplace* at his side. Laplace's present-day fame rests mainly on his mathematical and physical achievements. However, he was very broadly informed and, despite being a theoretician, had a flair for experiments, and his critical way of thinking was trained by the strict rigour of theoretical mechanics. All that must have been a great boon to the chemist partner.

Probably it was not by mere chance that the chemist looked for the company of the somewhat younger physicist. He must have known that the chemistry of burning can be revealed only together with its physics. That was probably the reason why they started with research into thermal processes. At first, they studied the changes of state. They revealed the fact, already indicated by Boyle, that evaporation of a liquid is enhanced by both higher temperatures and lower pressures. In addition, they realized that heat, if administered to a substance that is evaporating or melting, does not increase its temperature: the changes of state need a certain amount of latent heat (that was previously observed also by Black). A most important observation of theirs was that a substance might appear in all its three states if heated or cooled appropriately. We think that in this area of research their most basic achievement is the distinction between heat and temperature thus making heat a quantity which can be measured. By this they established the basic ideas and methods of calorimetry, which is still in use today.

They summarized these ideas in their co-authored work entitled *Mémoire sur la chaleur*. Free heat which can be measured by making use of a thermometer was distinguished by them from hidden heat which is contained by the bodies. Nowadays, the first one is called temperature the other one simply heat. Let two bodies of the same mass be of equal temperature (using a more modern nomenclature), still their content of heat might be different. This is so because their chaleur specific, translated as their specific heat, might differ. This quantity can be measured for mutually mixing substances. For example, warm water can be mixed with cold alcohol. Their train of thought was the following. Let T_{water} denote the temperature of water and T_{alc} that of alcohol prior of mixing, and T the final temperature of the mixture. The mass of water is m_{water} and that of alcohol m_{alc}. They multiplied the masses with the respective changes of temperature finding that the ratio of the two products is independent both of masses and temperatures. An algebraic expression makes this statement clearer,

$$\frac{m_{alc}(T - T_{alc})}{m_{water}(T_{water} - T)} = \text{constant} \tag{3.16}$$

The constant was found to vary from pair to pair of liquids. Thus a certain quantity can be associated with each of the liquids being called by them the specific heat of the liquid, c. The constant on right hand side is their ratio, viz.

$$\frac{m_{alc}(T - T_{alc})}{m_{water}(T_{water} - T)} = \frac{c_{water}}{c_{alc}} \tag{3.17}$$

With these notions and symbols at hand one can explain the idea of heat. If the temperature of a substance changed from T_2 to T_1 it means that an amount of heat, equal to $m(T_2 - T_1)$, was gained or lost depending on which of the two temperatures was the higher. This is like filling two vessels of different cross sections with water. If we want to have the level of water in both, more water is to be poured into the wider vessel. Now the role of the height of the water column is played by temperature, that of the amount of water by heat whereas the cross sections correspond to the heat capacities i.e. the product's specific heat times mass, cm.

    If we pour some liquid from one vessel to another we feel it natural that the amount of liquid is constant, that is the liquid is conserved, provided we do not spill even a drop and also prevent evaporation. Calorimetric experiments show that heat is similar to a liquid in this respect, it being an entity whose amount does not change when it is transferred from one substance to another if care is taken that no heat be lost to a third substance; that is, if the substances are thermally isolated from the environment. Under such conditions heat is conserved. Two theories of heat were known to Lavoisier and Laplace and both of them are in line with this basic requirement. One of them which looked more obvious for their age was the theory of caloric. Caloric was thought to be a fluid, a very finely grained substance, which fills all the bodies, a theory which was also accepted by Lavoisier in his earlier works. The other theory was that of thermal motion, which regarded heat to be the uninterrupted motion of the particles of the substance. This idea was not new; something similar had earlier been suggested by *Bacon, D. Bernoulli* and *Newton*. The novelty was in the consideration that the conservation of heat also follows from this theory. Particles in motion conserve their kinetic energy being called vis viva in those times.

Heat is either a substance or kinetic energy. The experiences drawn from calorimetric observations were in accordance with both of the ideas. Lavoisier and Laplace concluded that the two theories are phenomenologically equivalent.

The method of mixing can be used only in exceptional cases for the measurement of heat. Apart from non-miscible liquids, it is inappropriate, for example, to measure the heat that evolves when hydrogen burns, or to determine "animal heat", the heat production of metabolism, a problem much discussed in those years. Curiously enough, they did not realize that it is needless to mix the substances of different temperatures, thermal contact is sufficient. (Putting a warm copper block into cool water the two substances reach a common temperature after some time and, by knowing the masses and the initial temperatures of both, the ratio of heat capacities can be evaluated in the same way as at the method of mixing. Nowadays that is the most common method of calorimetry.)

Lavoisier and Laplace developed the ice calorimeter as a solution. Here the amount of heat developed within the calorimeter is measured through the amount of ice which gets melted during the process. The only quantity needed is the latent heat of fusion for unit mass of ice. The method can be regarded as isothermal (constant temperature) calorimetry since the temperature of ice does not change in the course of measurement. In contrast, the methods described above are adiabatic (constant heat) methods where thermal isolation prevents any loss of heat.

Posterity holds that Lavoisier's most important work about burning is his anti-phlogistic essay entitled *Réflexions sur le phlogistique*. Here the theory he refuted was something much more sophisticated than Stahl's simple but effective idea which a few years earlier he himself regarded to be a "most successful" explanation of certain features of burning. Stahl's successors tried to understand some more recent observations within the framework of this model. Lavoisier's refutations were based on measurements of mass and heat.

The calcination (oxidation) of metals is also burning, hence phlogiston leaves the metal according to Stahl. Consequently, any metal oxide is expected to be lighter than the metal but in reality it is always heavier. The improved phlogiston model assumed that "free fire" enters the burning body. Lavoisier answered that this "free fire" ought to be very heavy, an assumption that had not been verified by any mass determination. Moreover, if "free fire" gets into the metal when it is oxidized this process should be accompanied by absorption of heat in contrary to the calorimetric experiments which show that oxidation goes always with evolution of heat.

A subtler version of phlogiston theory regarded phlogiston to be "light matter". This theory accepted that both burning and metal calcination consist in composition with oxygen, but in the course of this process "light matter" leaves the burning substance. Lavoisier, with his good understanding of calorimetry, refuted this idea by showing that if this were true the specific heats of metals would be higher than those of oxides whereas measurements show the opposite.

Let us add to the above arguments several further ones formulated earlier: substances can burn only if oxygen is present and the weight increase of the substance having been burnt is proportional to the volume of the oxygen spent. Thus it is not just needless to assume the presence of phlogiston in every combustible substance, but we do know that the existence of such a fire substance is impossible. As it was formulated a few years after Lavoisier's death by an English follower of his: "*Combustion is a process by which certain substances decompose oxygen gas, absorb its base, and suffer its caloric to escape in the state of sensible heat.*"

I would not advise anyone to repeat this sentence at a chemistry exam. The author seems to be ignorant of or at least hesitant whether oxygen is an element, thus he writes about its decomposition attributing the combustion to only one of its component, which he calls its base. Moreover, he thinks that the heat of combustion is due to the caloric content of the oxygen. To be sure, the existence of caloric, a fluid which fills the bodies, was not alien to Lavoisier this being one of the two alternative models by which he could interpret calorimetric observations. He dismissed the theory of phlogiston because the escape of a component of every combustible material during burning was against his experiences. This, however, does not preclude the existence of the ubiquitous caloric.

In our current usage, we would say that Lavoisier refuted the existence of phlogiston as the model of oxidation–reduction processes, but did not exclude the existence of caloric as the model of heat transfer processes.

4 The Balloon

For centuries people believed that it is easier to make gold than to fly. By the end of the eighteenth century what has been proven was the opposite of this. Once Black wanted to demonstrate the low density of hydrogen to his guests in a spectacular manner. He took a thin animal bladder, a calf's caul, filled it with hydrogen and let it loose. The bladder flew as high as the ceiling. First the guests suspected that someone pulled it with a thin thread from the upper floor, since anything can move only if it is moved. Only after having touched this primordial nursery toy did they put up with the need for a different explanation.

It was in these time that two well-to-do paper manufacturers, the *Montgolfier* brothers set out to perform balloon experiments. Their first successful flights, first without any payload, then with animals, finally with two passengers, were performed through the buoyant force of hot air. A huge linen sack lined with paper, under which a huge fire of chaff was made, spent many minutes in the air covering several miles from the site of ascent due to the favour of a lucky breeze.

Physicist *Charles* took a different path making use of his knowledge and experience regarding the physics of gases. He made his balloon of fine silk and made it gas tight impregnating it with a rubber solution. The sack was inflated into a sphere of a diameter of 13 feet with hydrogen which was developed at the site of ascent (Fig. 6). 1000 pounds of iron filing was reacted

Fig. 6 Flight of Charles's balloon (from Wikimedia Commons, figure in public domain)

with 500 pounds of sulphuric acid in a matter of several days until the gas filled the sack. Then it was able to float 45 min, to cover 15 miles and as it descended frightened the nearby people that much that they tore the mysterious contrivance into pieces.

Balloon experiments, in parallel with hot air and hydrogen, were continued. From our present point of view, it is important as the first practical use of hydrogen gas. *Benjamin Franklin*, however, who stayed in Paris in those days, having been present at the flight was most optimistic about the invention. "*The invention of the balloon appears, as you observe, to be a discovery of great importance. Convincing sovereigns of the folly of wars may perhaps be one effect of it, since it will be impossible for the most potent of them to guard his dominions. Five thousand balloons, capable of raising two men each, could not cost more than five ships of the line; and where is the prince who can afford so to cover his country with troops for its defence as that 10,000 men descending from the clouds might not in many places do an infinite deal of mischief before a force could be brought together to repel them?*" That was what he wrote.

Franklin was a great man. We, however, having the experiences of two and a half centuries since his time, would like to have different foundations for optimism.

Further Reading

J.R. Partington, *A History of Chemistry Vol. I–IV* (MacMillan, London, 1961–1970)

D.M. Knight (arr., introd.), *Classical Scientific Papers—Chemistry* (Mills and Boon Limited, London, 1970)

M.J. Nye, *The Question of the Atom*. Volume IV of the Series *The History of Modern Physics 1800–1955* (Tomash Publishers, Los Angeles, 1984)

R. Schiller, *Between One Culture* (Springer Nature Switzerland AG, 2019)

Between Chemistry and Physics

1   The Well-Chosen Unit of Mass

All that was assumed, argued about and comprehended by the chemists of the nineteenth century has become settled as secondary school curriculum by now. As we reach fifteen years of age all of us would know what the difference is between mixture and compound, what the meaning is of a chemical formula, or of atomic mass. People are familiar with these notions which have been household words for a long time. No one has ever met a person who had but the faintest doubt of their meaning or truth. Thus, it is understandable that their origin, the way in which this part of knowledge was accrued, went into oblivion. Usually we quote the laws of definite proportions and multiple proportions (that is, the weight proportions of the constituent atoms in a compound) as the bases of classical chemical thinking. These have rested on the simple, quantitative observations of chemists who dealt with material synthesis and analysis, who accepted the conservation of matter as a basic law and whose most important, sometimes only instrument was the balance.

Elements and compounds are discerned through measurements of weight. Elements are substances which can be transformed into heavier substances only, whatever chemical reaction they undergo. For example, hydrogen is an element since out of 10 g of hydrogen one obtains 90 g of water after its reaction with oxygen. No chemical reaction exists which, starting with 10 g of hydrogen, would yield some hydrogen containing substance that weighs less than 10 g.

© The Author(s), under exclusive license to Springer Nature
Switzerland AG 2022
R. Schiller, *A Non-Traditional Guide to Physical Chemistry*,
https://doi.org/10.1007/978-3-031-07488-2_2

The law of definite proportions was expressed by *Proust* in its clearest, classical form at the turn of the 18th to nineteenth century. He stated that, irrespective of how a compound was prepared, the weight proportions of its constituents are always the same. This law, tacitly though, had been accepted by the chemists much earlier since, without supposing that it is true, there would have been no sense to perform any quantitative analysis. However, the exact measurements and the clear formulation of the law are Proust's merit. First he described the composition of the native copper compound, malachite, a basic copper carbonate, the composition of which is given now as $CuCO_3Cu(OH)_2$. He wrote "*One hundred pounds of copper, dissolved in sulphuric or nitric acids and precipitated by the carbonates of soda or potash gives 180 pounds of green carbonate. If this quantity be submitted to a gradual distillation it gives 10 pounds of water[…] this water only passes over successively and conjointly with the acid* (carbon dioxide). *Deprived of these two components the carbonate leaves 125 pounds of black oxide at the bottom of the retort. […] One may, in all analysis, take 180 pounds of carbonate or 125 pounds of black oxide for a quintal* (112 pounds) *of copper. Native carbonates of copper are also found in this ratio of oxidation.*" The gist of the passage is given in the last sentence.

Not everyone agreed. *Berthollet* wrote the following: "*Chemists, struck by the fact that they find definite proportions in many combinations, often consider it a general property of combinations that they are composed of definite proportions […]*" and cited a number of counter examples. It seems obvious to us that this great chemist did not always make a difference between compound and mixture. Proust, however, could make a clear distinction and having been involved in an argument with Berthollet, expressed his stance clearly: "*These ever-invariable proportions, these constant attributes [..] characterize true compounds of art or of nature [..].*"

Proust being an experimentalist based his statement on laboratory observations. *Dalton*, his contemporary, was directed to the law of multiple proportions by the atomic hypothesis. The law refers to compounds which consist of the same elements but their proportions are different in each compound. Such families of compounds are, for example, carbon monoxide and carbon dioxide, water and hydrogen peroxide, or the six oxides of chlorine. If one assumes, in accordance with all the atomists since *Democritus*, that any material consists in different combinations of atoms, then there is only one further step missing. Dalton made this, as it was described by a contemporary of his. "*The hypothesis upon which the whole of Mr. Dalton's notions respecting chemical elements is founded, is this: When two elements unite to form a third substance, it is to be presumed that one atom of one joins to one atom of the other, unless when some reason can be assigned for supposing the contrary. [...] Whenever more than one compound is formed in the combination of two elements, then the next simple combination must, he supposes, arise from the union of one atom of one with two atoms of the other.*"

Indeed, in compounds which are built by the same elements, the relative masses of the elements are given by small integers. Take for example carbon oxides. Starting with 1 g carbon one gets either 3.67 g carbon dioxide or 2.31 g carbon monoxide, i.e. the mass of oxygen in carbon dioxide is 2.67 g, in carbon monoxide 1.33 g for 1 g of carbon. The ratio of the oxygen contents of the two compounds is close to 2.

The atomic hypothesis and the law of multiple proportions are seen to be in apparent harmony. The figure taken from Dalton's book, published in 1808 (Fig. 1), clearly shows the strength of the simple, suggestive and still live model the author has given us. Nevertheless, some readers of this book maintain that the symbols used hark back to the times of the alchemists. As it is seen, atoms are represented as tiny spheres which form molecules by coupling with each other. The power of intuition is striking; the idea, however, was much in want of proof.

Although knowing nothing more of the compositions than their weight Dalton tried to determine the atomic masses. He took the mass of one hydrogen atom as the unit of mass, believing that the lightest substance (the one of the lowest density) must be built by the lightest atoms. This suggestion seems to be somewhat naïve: light atoms might form a high density substance if they are closely packed. In water the mass of oxygen is eight times the mass of hydrogen. Thus, he thought that the atomic mass of oxygen is 8, since(!) it is known that the composition of water is OH, this being the simplest combination of the two atoms.

In light of our present knowledge, the error is obvious because hydrogen gas is known to consist of diatomic molecules, H_2. Consequently, the composition of water is H_2O and the atomic mass of oxygen, taking the mass of the hydrogen atom the unit, is about 16. All this, however, cannot be revealed through the law of multiple proportions.

The determination of the atomic masses rests on Dalton's assumption which ascertains: "*When only one combination of two bodies can be obtained, it must be assumed to be a binary one, unless some cause appear to the contrary.*" That sentence conjures the spirit of the Middle Ages up although we are taught that this is the start of modern chemistry. "*Entities are not to be multiplied beyond necessity.*" That was said by the Franciscan scholastic theologian, William Ockham, some time in the first half of the fourteenth century. Later an apocryphal version of the sentence was called Ockham's razor. As understood by Bertrand Russel, it is not reasonable to assume an entity which is

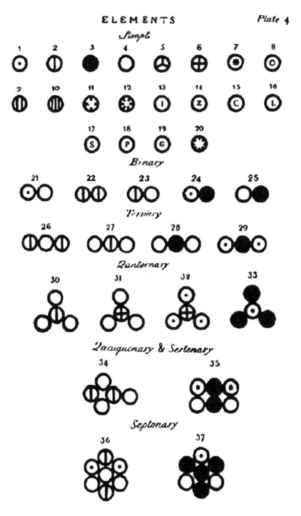

Fig. 1 Atoms and molecules in Dalton's book

needless to explain a scientific result. Dalton's above quotation shaved all the molecules which are built of two elements and more than two atoms with Ockham's often useful but still dangerous tool. Not seeing any cause to the contrary, he stated that the composition of the simplest compound made of hydrogen and oxygen is OH. The simplest compound of carbon and oxygen is CO. Today the first formula is known to be false, the second to be correct.

As far as algebra is concerned, the problem is simple. Simply hopeless. The mass balance of the reaction $A + B = AB$, by knowing mass conservation law, is $m_A + m_B = m_{AB}$. According to atomic theory the masses are $m_A = n_A M_A$, $m_B = n_B M_B$, and $m_{AB} = n_{AB} M_{AB}$ where n_A...

denote the number of atoms or molecules and M_A ... the atomic or molecular masses. One can measure only masses m with a balance. As far as n values are concerned our imaginations are set free. There are six unknown quantities but only four equations. This was clear also to Dalton and that was the reason why he had to take recourse to the combination of atoms and the assumption of the composition of the simplest molecules.

The analysis of the composition by weight cannot reveal the number of atoms in the sample. Some additional source of information is needed. At the beginning of the nineteenth century a simple law was found for the volumes of chemically reacting gases by *Alexander von Humboldt* and *Gay-Lussac*. As they observed, for example, that a certain volume of ammonia and the same volume of gaseous hydrochloric acid totally neutralize each other; or two volume of hydrogen and one volume of oxygen form water, consuming both gases completely. Gay-Lussac summarized these and similar observations the following way: "*I have shown [...] that the compounds of gaseous substances with each other are always formed in very simple ratios, so that representing one of them by unity, the other is 1, or 2, or at most 3. These ratios by volume are not observed with solid or liquid substances, nor when we consider weights, and they form a new proof that it is only in the gaseous state that substances are in the same circumstances and obey regular laws.*" Most naturally, these strict proportions hold only, if the reacting gases are of the same temperature and pressure.

The atomic hypothesis, together with the above observations on the volume proportions of reacting gases, directed *Avogadro* to a new assumption. "*M. Gay-Lussac has shown in an interesting Memoir that gases always unite in a very simple proportion of volume, and that when the result of the union is gas, its volume also is very simply related to those of its components. But the quantitative proportions of substances in compounds seem only to depend on the relative number of molecules which combine, and on the number of composite molecules which result. It must be then admitted that very simple relations also exist between the volumes of gaseous substances and the numbers of simple or compound molecules which form them. The first hypothesis to present itself in this connection, and apparently even the only admissible one, is the supposition that the number of integral molecules in any gases is always the same for equal volumes, or always proportional to the volume.*" (Chemical nomenclature was different from ours in Avogadro's time. He calls *molecule* both atoms and molecules, *elementary molecule* is an atom, whereas *constituent molecule* denotes the molecule of an element, and *integral molecule* that of a compound).

The present-day formulation of the hypothesis has become more exact through the stipulation that volumes must be taken at the same pressure and temperature—something what Avogadro tacitly assumed. Thus, as far as gases are concerned, Dalton's conjecture turned out to be correct: the lighter a substance is, the lighter its constituent atoms are.

Now, the analyses by weight, together with Avogadro's assumption, are sufficient to determine the relative atomic masses; that is what is called atomic mass. Weight determinations render the sum of the mass of the reacting molecules; Avogadro's hypothesis tells the relative change of the number of molecules in the course of the gas reaction, since volumes are thought to be proportional to the number of molecules. Volume, V and number of molecules, n being proportional there hold the simple relations $V_A/V_B = n_A/n_B$ and $V_A/V_{AB} = n_A/n_{AB}$. By knowing the sum of the masses of the reacting molecules, together with their relative volumes one can evaluate the relative molecular masses. Two equations were missing for Dalton to solve the algebraic problem, Gay-Lussac rendered the above two.

Thus, everything is revealed? Are the basic problems of stoichiometry solved? So it seems. (Stoichiometry: this is the name of the calculations on the relative quantities of reacting substances. An example of a stoichiometric equation is the one in the previous chapter describing the burning of phosphorous.) Just by making use of his own simple hypothesis Avogadro corrected some of Dalton's statements: he showed that the elementary gas molecules are diatomic: H_2, O_2 etc., the composition of water is H_2O, that of ammonia is NH_3. Indeed, these are the last words on this matter, we do not think differently even today. The classical molecular mass determinations, important methods developed by *Dumas*, are all based on the ideas of Avogadro and Gay-Lussac.

It seems to be essential to choose the mass of an atom as the unit of atomic masses, possibly that of the lightest one, lest no number less than one occur. Thus the H atom mass was chosen. Nowadays, a somewhat different unit is in use, as it will be set forth later.

Ideas and methods had been clarified only at a slow pace. It took almost 60 years after the work of Dalton and Avogadro for the notions of atom and molecule to obtain an unequivocal definition and the above ideas on the determination of atomic and molecular masses to become generally accepted. A most important international chemical congress was held in Karlsruhe in September 1860. In order the aim of its convener, *Kekulé*, was to clarify these questions. His ambition must have seemed modest, he wanted to make the chemists agree regarding the notions of "atom", "molecule" and "combining weight" and to agree on names and symbols. An obvious precondition of all these must have been an agreement on the structural model of chemical substances. The organizers were cautious since they knew that it was impossible to eliminate all the differences at once. Nevertheless, their aim was clear:

"*We are assembled for the specific goal of attempting to initiate unification around points of vital concern for our beautiful science.*"

The discussions, however, bore but little fruit. Finally even Kekulé, a devoted atomist himself, doubted whether the notion "atom" is of the same meaning for a chemist and a physicist: "*The chemical molecules have not been shown to be identical with gaseous molecules. Thus it is not established if the smallest quantity of a substance that enters into a reaction is also the smallest quantity of this substance that plays a role in heat phenomena.*"

One of the attendees of the congress, *Cannizzaro*, distributed the copies of an essay of his, written two years earlier. This text turned out to be the last word in the debate, giving the order of the notions in a way that is taken as valid even today. It could be copied into a present-day introductory textbook. "*[...] the weights of the molecules are proportional to the densities of the substances in the gaseous state. If we wish the densities of vapours to express the weights of the molecules, it is expedient to refer them all to the density of a simple gas taken as unity. [...] Hydrogen being the lightest gas, we may take it as the unit which we refer the densities of other gaseous bodies, which in such a case express the weights of the molecules compared to the weight of the molecule of hydrogen. [...] Since I prefer to take as a common unit for the weights of the molecules and for their fractions, the weight of a half and not of a whole molecule of hydrogen, I therefore refer the densities of the various gaseous bodies to that of hydrogen = 2.*"

This choice being most practicable is tantamount to the understanding that the hydrogen gas is made of diatomic molecules, H_2. This agrees, for example, with the formation of gaseous hydrochloric acid from the elements, as

$$\text{hydrogen} + \text{chlorine} = \text{hydrochloric acid} \qquad (1.1)$$

a process the mass balance of which is

$$2 \text{ mass unit hydrogen} + 35.5 \text{ mass unit chlorine}$$
$$= 37.5 \text{ mass unit hydrochloric acid} \qquad (1.2)$$

whereas the volume ratios are given as

$$1 \text{ volume unit hydrogen} + 1 \text{ volume unit chlorine}$$
$$= 2 \text{ volume unit hydrochloric acid} \qquad (1.3)$$

The atomic mass of an element is equal to its mass in one volume unit, i.e. to its density in the gaseous state. Expressed in the more usual way, the atomic mass unit is the mass of the hydrogen atom. We used to be taught that the valence of an element equals the number of hydrogen atoms in a compound. That means only that when the element combines with hydrogen in the gas phase its valence equals the number of volume units of hydrogen consumed. Thus the valence of chlorine is 1, that of oxygen is 2.

If an element has no known hydrogen compound, then both atomic mass and valence must be determined in some indirect way. Let us take copper, for example, which lacks any stable compound with hydrogen. However, analyzing the masses of the constituents in copper chloride, one finds that for 71 g chlorine 63.54 g copper is present. From the studies with hydrochloric acid one also knows that 71 g of chlorine combines with 2 g of hydrogen. Hence, copper is equivalent with two atoms of hydrogen, its valence equals 2 in that compound.

That was the way our present-day concepts of atomic and molecular masses were attained already around the middle of the nineteenth century. But it should not be forgotten that all this was achieved on the basis of two hypotheses, assumptions which did not have any direct experimental proof. As far as some direct proof for the discontinuous, atomic structure of matter was concerned, Dalton or Cannizzaro was in no better position than *Leucippus* or *Democritus* had been. There is little doubt that many of the scattered observations of the chemists were successfully organized in terms of not too complex ideas. But the sheer fact that a theory is based on reasonable assumptions, is not self-contradictory and appropriately describes a part of the observations which cannot be regarded as a proof.

By hindsight, considering the formation of the chemists' ways of thinking, one might be prone to believe that the nineteenth century development of chemistry was a triumph of atomic theory which preceded physical knowledge and proofs in this area. But this is an obviously absurd opinion, since no serious theory of matter can be envisaged without any solid knowledge of physics. On the contrary, soon it will be shown that as physical chemistry had become a valid branch of science, atomic hypothesis was refuted at first and it happened only much later that the existence of atoms was proven.

Atomic theory was the subject matter of the experimenting chemists, and not even all of them. *Faraday*, one of the most brilliant minds, who was said "to smell the truth", did not believe in the existence of molecules built from atoms in the way as Dalton had proposed. Also one might find it characteristic of the period that in the years after the Karlsruhe Congress *Brodie*, a critically minded professor of chemistry, tried to develop a mathematical system, independent of the atomic hypothesis, for the description of the chemical experiences.

Still chemist-experimentalists could find a number of convincing arguments in favour of the atomic hypothesis by the second half of the nineteenth century. These were not based on the physical properties of atoms or molecules, much more on the harmony between chemical observations and the material model of tiny balls. To be more exact, the discontinuous model was refined so as to achieve an accordance with the experimental results.

The name of *Wöhler* often appears in the introductory chapters of organic chemistry textbooks because he was the first chemist who succeeded in synthetizing an organic compound from an inorganic one in vitro; that is, in his laboratory. Thus, he disproved the earlier belief that only a living organism is able to produce organic compounds. Heating ammonium cyanate, an inorganic compound he obtained the well-known organic compound, urea. The proportions of the constituents are the same in the two compounds (now we know that also their molecular masses are equal) their properties, however, are very different. Their empirical formula—this the name of the expression which shows nothing else but the number of atoms in a molecule—is the same, CN_2H_4O. Still their physical and chemical properties differ. This phenomenon is called isomerism, and the molecules of identical empirical formulae but of different properties are each other's isomers. Wöhler only realized the importance of his discovery in this aspect, but he remained cautious: "*I refrain from all the considerations which so naturally offer themselves, particularly those bearing upon the composition relations of organic substances, upon the like elementary and quantitative composition of very different properties.*"

Several decades later Kékulé could make a bolder step forward. Having constructed the notion of valence bond (as we call it nowadays, he called it affinity unit) he introduced the idea of the structural formula. Valence meant for him something more than relative volumes or weights. From this time on, the fact that carbon is tetravalent—this was also established by him—has meant something more than the analytical fact that 12 g carbon can combine with 4 g hydrogen.

According to his concept that also means that one atom C can bind four atom H forming one molecule of "marsh gas", methane of the formula CH_4. If so, it seems reasonable to consider the relative order of the atoms. Open carbon chain compounds consisting only of C and H, called alkanes, are built from a chain of CH_2 groups with a CH_3 group at each end, that way,

$$CH_3-CH_2-CH_2 \cdots CH_2-CH_3$$

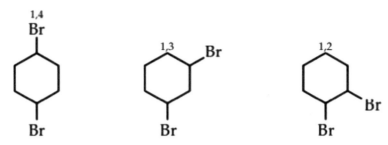

Fig. 2 Two H atoms are replaced by two Br atoms. Three isomers of di-brome benzene are seen to exist. The figure, drawn by Körner in 1874, is very similar to the present-day symbols

The composition of benzene, however, must have been understood by a much different atomic order. The solution is well known, it is a ring structure that complies with the empirical formula C_6H_6,

$$
\begin{array}{c}
CH = CH \\
\diagup \qquad \diagdown \\
CH \qquad\qquad CH \\
\diagdown\!\!\diagdown \qquad \diagup\!\!\diagup \\
CH - CH
\end{array}
$$

Kekulé's great achievement was not just to find out the structural formula of benzene. A much more central insight was that the sequence of atoms also counts; in other words, that structural formulae are to be constructed and considered.

Körner, a student of his, following these lines investigated the isomers of substituted benzenes, that is compounds where one or several H atoms are replaced by some other atoms, for example by Br. Why do three different dibromobenzenes exist of the common empirical formula $C_6H_4Br_2$ and again three different tribromobenzenes, $C_6H_3Br_3$? The answer is given in Fig. 2 taken from his original paper: the relative positions of the bromine atom are different in each of the molecules.

This was, however, only the beginning of the story. Drawing the structures of a number of organic compounds, *van't Hoff* came to the conclusion that there are cases where the number of isomers on paper outdoes that of the compounds in the alembic. For example, considering the derivatives of methane, one can draw only one isomer of chloromethane, CH_3Cl in accordance with the experiments, but two isomers of dichloromethane, since the two Cl atoms are either in a neighbouring or a diagonal position:

$$
\begin{array}{ccc}
\overset{\displaystyle Cl}{\underset{\displaystyle H}{H-C-Cl}} & & \overset{\displaystyle Cl}{\underset{\displaystyle Cl}{H-C-H}}
\end{array}
$$

The experimentalist, however, knows only one compound of this composition. The compound CHClBrI should have three isomers, as it is seen below, still there exist only two.

$$
\begin{array}{ccc}
\overset{\displaystyle I}{\underset{\displaystyle H}{Cl-C-Br}} & \overset{\displaystyle I}{\underset{\displaystyle Br}{H-C-Cl}} & \overset{\displaystyle I}{\underset{\displaystyle Cl}{H-C-Br}}
\end{array}
$$

These sort of contradictions can be solved if the molecules are thought of not as planar but rather as three-dimensional spatial structures. According to van't Hoff: "*The theory is brought into accord with the facts if we consider the affinities* [valences] *of the carbon atom directed toward the corners of a tetrahedron of which the carbon atom itself occupies the centre.*"

Figure 3 is from van't Hoff's paper, Fig. 4 from any of a modern organic chemistry textbook. Let us imagine two Cl atoms to any two corners of the tetrahedron, soon we can be convinced that only one dichlorobenzene can be constructed in accordance with the experimental fact.

Compounds, similar to CHClBrI in the sense that different atoms or atomic groups are bound by each valence of the carbon, exist in the form of two isomers; this being a consequence of the tetrahedral valence orientation. The peculiarity of such pairs of isomers, as they are observed, is their complete identity in all of their physical and chemical properties save but one, and this is their response to polarized light.

Polarized light is a light wave which oscillates only one single direction (Fig. 5). There exist certain crystals which, if polar light goes through them, its plain of oscillation (polarization plain) gets rotated. Such crystals occur always in pairs, and they are each other's mirror images (Fig. 6). Being called asymmetric crystals they make the polarization plain rotate to the same extent but of opposite direction. In short they are optically active.

Having investigated both the solid crystals and the aqueous solutions of tartaric acid, *Pasteur* made clear that solutions can act on polarized light in the same way that crystals do. Also solutions can be optically active. Considering the analogy between crystalline and molecular symmetries Pasteur, van't

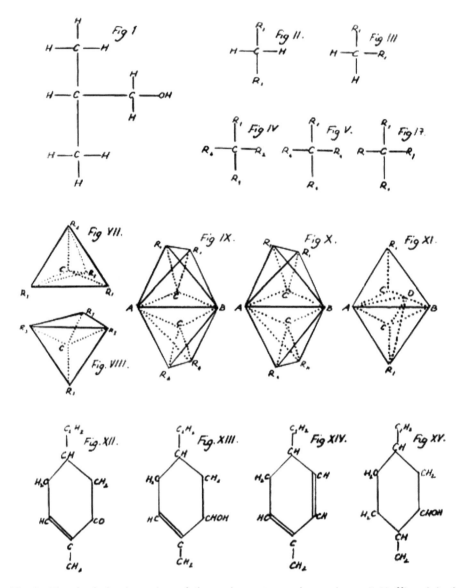

Fig. 3 Tetrahedral orientation of the carbon atom valences in van't Hoff's original work

Hoff and *Le Bel* came to the conclusion that the phenomenon must be ascribed to the asymmetry of molecules. The picture of a tetrahedral carbon atom was in complete agreement with this idea. A tetrahedron with different atoms or groups at each of its corner is asymmetric in the same sense as an optically active crystal, also can have two forms these being each other's mirror images. The two forms differ in the order of the atoms at the corners.

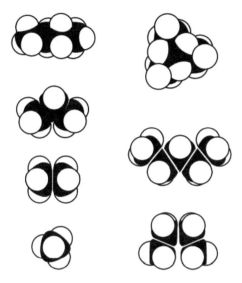

Fig. 4 Molecules built of tetrahedral carbon atoms according to our present-day visualization

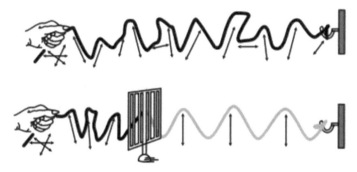

Fig. 5 Natural light is visualized through a string swinging in every direction. Polarized light is similar to a string which swings only one direction

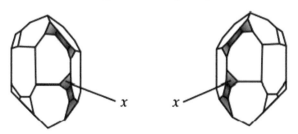

Fig. 6 Two quartz scrystals are each other's mirror images. If polar light traverses one of them its plane of polarization changes. Traversing thereafter the second crystal the original plane is restored

The appearance of the structural formulae, the understanding of isomerism, the realization of the tetrahedral symmetry of carbon had a bearing more important than just the interpretation of certain chemical phenomena. In the course of chemical research over decades the atoms appeared to be more and more real, they were thought to be almost tangible entities with well-defined mechanical properties. It turned out that chemical reality is in good consonance with a number of consequences of the mechanical atomic and molecular models. Chemists had less and less motive to doubt the existence of atoms.

Still, in the second half of the nineteenth century atomic theory had no physical foundation of mathematical rigour. One sees drawings, reads reasonable arguments in the chemistry papers of the age; one would look for physical measurements or mathematical arguments in vain. Although it was the age of *Hamilton* and *Maxwell*, the birth of physical theories with high mathematical demand. The construction of molecules or the essence of the valence bond was not tried to be treated in a way similar to the strict theories of physics. Exactitude appeared in other areas of chemical thinking.

2   Atoms Counted and Weighed

"Have you ever seen an atom"? This is the archetype of stupid questions. Stupid, but not only because it is obviously impossible to see an atom directly, with the naked eye, in the sense as we see Jill in the neighbour's garden. There are, of course, instruments which by making use of some combination of physical laws and resourceful engineering inform us about certain facts in terms of flickering numbers or perhaps of a figure on a graphic display. All this, however, has little to do with "seeing". But this only shows the question to be ignorant, not stupid. Its stupidity consists in the suggestion that the existence of an object can be taken for granted if it can be seen, perhaps even touched. Scientific thinking, on the contrary, becomes more and more certain about the existence of an entity if its properties, as many of them as possible, can be determined by independent methods. For example, we can rest assured that atoms exist if we can determine their quantity or absolute mass independent of chemical experiences.

The number of atoms was determined by *Loschmidt* in a simple way around the middle of the nineteenth century. He regarded gas atoms as small balls the radii of which are negligible in comparison to the distances between the atoms. If the gas is liquefied, the atoms get close to each other filling almost completely the space. The liquid volume equals the volume of an atom multiplied by their number, N,

$$V_{\text{liquid}} = N \frac{4\pi}{3} r^3 \qquad (2.1)$$

where r denotes the radius of an atom.

Thus, the liquid volume renders the product of the number and volume of the atoms according to this estimation. A further relationship is needed for the separation of the two quantities. Here the motion of the atoms in the gaseous state must be considered.

Atoms of a gas make a thin population, yet, despite their large mutual distances, there must be some interaction between them. The experiment given in Fig. 7 is a proof of that. There are two conveyors belts in a common gas space; one of them being driven the other one will be set in motion some time after the first; both of them run in the same direction although no force seems to act on the second belt.

The phenomenon can be explained through the motion of the gas atoms. The atoms flying at random hit the driven belt and get accelerated in the direction of its motion. Moving on they reach the steady belt and impart their own velocity to it. Since this velocity is greater in the direction of the driven belt's motion they make the steady belt move in the same sense.

This observation and explanation point to the existence of the gas atoms but do not indicate any interaction between them. A further experiment helps. The farther the two belts are from each other, the more sluggish the response of

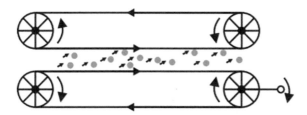

Fig. 7 A proof for the mutual collisions of the particles of a gas (After F. de Körösy: *An Approach to Chemistry*, Pitman, London 1969, with the permission of the author). Two conveyor belts are in a common gas space. One of them being driven sets in motion the other one; the smaller the distance between the belts, the faster it moves. The driven belt accelerates the gas particles in the direction of its own motion and the particles set the other belt into motion—the nearer the two belts are the more effectively this happens

the non-driven belt. This can be explained through the interaction of atoms. The accelerated atoms, as they fly off the driven belt, collide with other atoms, thus giving them their extra velocity. These latter ones, though now faster than before, might fly in any direction, the particular direction given by the belt "gets forgotten" in the course of atomic encounters.

That is, however, an explanation not the experience, this latter being only that though the force is transmitted by the gas it does not happen without any hindrance. The gas layer in close proximity with the driven belt moves together with it whereas the further portions move slower. This phenomenon is called viscosity. It is a ubiquitous effect which, for example, shows in the way a gas or liquid flows in a tube: it flows the fastest in the middle whereas it is at rest at the wall. This is depicted in Fig. 8.

That experiment and some similar ones might take us to the frequency of atom–atom collisions, or what amounts to the same thing, the average distance covered by an atom between two encounters. The zig-zag path of a particle is seen in Fig. 9; an atom changes its direction when it hits another, similarly to the billiard balls on the table. Obviously, by the larger cross sections of the balls and by being more of them in a given volume, their encounters become more frequent, the straight path between two successive encounters becomes shorter.

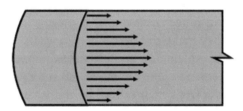

Fig. 8 Viscous flow of a gas or liquid. The arrow lengths are proportional to the fluid velocities

Fig. 9 Path of a gas atom; the particle flies along a straight line between two collisions

The straight portion between two breaks is called free path, it may be shorter or longer as it is controlled by chance. The average of these lengths, called average mean free path depends only on the number and size of the atoms. According to the picture of collisions, it seems easy to visualize that the mean free path is inversely proportional to the product of the number and cross section of the particles,

$$\text{mean free path} \propto \frac{1}{Nr^2} \qquad (2.2)$$

Collisions inform us on r^2, volume of the liquid state on Nr^3, two equations with two unknown quantities—fine! Loschmidt performed the simple calculation for a number of gases finding that different gases contain the same amount of particles in masses which are proportional to their atomic or molar masses: 2 g of hydrogen, 32 g of oxygen, 28 g of nitrogen consist of the same amount of molecules. In short, the conjecture of Avogadro and Cannizzaro has been substantiated.

According to modern measurements, one mole of a substance consists of $N_A = 6.023 \times 10^{23}$ particles/mole. (One mole of an element or compound is the mass expressed in grams which is equal with the atomic or molar mass of the substance.) N_A is called Avogadro constant. At present it is regarded as a defined quantity.

The result was deduced from the behaviour of gases but clearly it refers to all states. Let a substance be liquefied or frozen, neither its mass nor the number of its particles would change.

All this being known, the chemist expresses the amount of a substance not in terms of mass or volume but of the number of moles. This is the natural measure of the amount of a substance because it is proportional to the number of molecules in a substance. Thus, the measurement goes back to the primordial determination of an amount, i.e. to counting, as it was done by our distant ancestors who gave their fortune of goats or stone axes this way. The only source of confusion may be the unit used because it is not 1 but a huge bunch of entities, the Avogadro constant. "Mole is like the score" that was how an engineer-scientist once understood the concept.

If both the weight of one mole of substance called the molar weight, M and the Avogadro constant are known, the molecular mass, m can be obtained through a simple division,

$$m = \frac{M}{N_A}. \qquad (2.3)$$

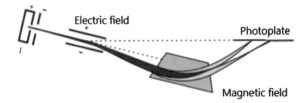

Fig. 10 Scheme of a mass spectrometer

The concept of the atomic, discontinuous structure of matter can be made more convincing if the molecular mass, hence the Avogadro constant is determined in a way which is independent of the above reasoning. The method that enables one to determine m directly is so simple and pictorial that we may have the feeling that the molecules are weighed on a balance one by one.

An instrument called mass spectrometer is devised to do this job. The basic idea is that first the atoms or molecules obtain electric charges i.e. get ionized. Both electric and magnetic fields act on moving charged bodies changing their motion. The electric field accelerates them, while the magnetic field deflects their path from the straight line. A light body is easier to be distorted in its flight than a heavy one hence the paths of the particles differ according to differences in masses. The scheme of the equipment is given in Fig. 10.

Fast electrons traverse the low-pressure gas in the ionization chamber, I. The electrons hit some of the gas molecules ejecting one electron of charge $-e$ from each, leaving the molecule with the positive charge $+e$. The charged molecules get accelerated by an electric field of voltage, U. Since, loosely speaking, voltage means the potential energy of a body of unit charge, eU is the ion's potential energy at the site of its formation. This is transformed into the kinetic energy of the ion, accelerated to its final velocity, v,

$$eU = \frac{1}{2}mv^2. \tag{2.4}$$

Magnetic field effects only moving charges. The magnetic force, F is perpendicular both to the direction of the field and the motion of the charge. Denoting the magnetic field strength by H the force is given by

$$F = ev_\perp H \tag{2.5}$$

Here $v_\perp$ denotes the velocity component of the particle perpendicular to the magnetic field.

Magnetic field cannot do any work on the charged particle because force and displacement are always perpendicular. Hence the ion traverses the magnetic field with constant speed, however, its direction changes. The magnetic field, H in Fig. 10 is the same in every point (it is said to be homogeneous) and is everywhere perpendicular to v. All this compels the ion to fly along a circular orbit, the deflecting force acting radially. The centrifugal force of a mass point moving along a circular path is mv^2/R and this is equal with the deflecting magnetic force,

$$evH = \frac{mv^2}{R} \tag{2.6}$$

The above equations define the radius of the circular path as,

$$R = \frac{\sqrt{2U}}{H} \sqrt{\frac{m}{e}} \tag{2.7}$$

The radius depends on the ratio of the charge and mass of the ion, e/m; this is called the ion's specific charge. The larger m, the larger R as one would expect: heavier particles are more difficult to deflect. Particles of identical charges but of different masses fly along different paths similarly to the components of white light as they are decomposed if the light shines through a glass prism (this similarity is referred to by the word spectrometry in the name of the instrument). By knowing the geometry of the equipment R can be determined and since both U and H is known, m/e is easy to evaluate. If one wants to know m so e must be measured by some independent method, which is to be given soon.

Mass spectrometry renders the individual masses of charged atoms or molecules one by one through measuring macroscopic quantities, R, U and H. These masses and molar masses again give Avogadro constant a value which agrees reasonably well with the data obtained by other, independent methods. The mass determinations verify the atomic hypothesis of the chemists, an idea the sole basis of which was the intuition of Dalton and Avogadro.

A tacit assumption was made in the description of the mass spectrometer. Electricity was thought to have atomic structure, since it was taken for granted that the same amount of negative charge is removed from each and every molecule. However, no care was taken to ensure this. If it happens that every molecule obtains the same amount of positive charge, $+e$ this points to the fact that this charge is the smallest, indivisible element of electricity. This reasoning alone, however, is not sufficient; the atomic structure of electricity

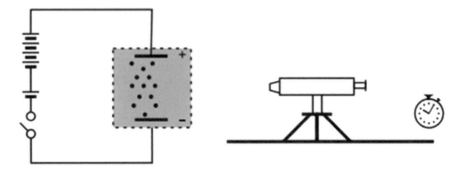

Fig. 11 The Millikan experiment

is supported by independent experiments, the existence of the elementary charge has been proven, its value has been measured.

Millikan's method resembles mass spectrometry as far as the basic idea goes, since both of them is based on mechanical forces acting on electrically charged bodies. A plate condenser, two parallel metal plates, is seen in Fig. 11, a voltage being switched to the plates. An electrically charged oil drop is floating in the air between the plates. It cannot drop because the sign and magnitude of the electric field is chosen so as to compensate for the gravitational force. The heavier the drop is, the stronger the field necessary and the more charge is carried by it, the weaker the field sufficient. Hence by knowing the electric field strength and the mass, m of the drop being afloat the amount of charge can be evaluated.

Electric field strength, E is the force which acts on the unit charge. The drop carrying charge q is exposed to the electric force qE which, if the sign of E is properly chosen, acts against the gravitational force, $m_d g$, where m_d denotes the mass of the drop and g the acceleration constant of gravity. The drop floats if the two opposite forces are equal,

$$m_d g = q E \tag{2.8}$$

The drop is usually charged through friction, whereas the amount of charge can be altered by illuminating the drop with ultraviolet light or by exposing it to radioactive radiation. Repeating the experiment with the same drop, which carries different charges, then again with different drops one finds invariably, that q is always the multiple of a constant, $q = ne$, where n is a small integer. The charge is never smaller than e. This is the indivisible elementary charge, the atom of electricity. According to contemporary measurements this elementary charge equals $e = 4.802 \times 10^{-10}$ electrostatic unit, or 1.602×10^{-19} C.

This is the smallest free charge which exists in nature. That is the charge of an electron and this is the amount of the positive charge which is carried by an atom after it has been deprived of one of its electrons.

By mass spectrometry one measures the quotient of ion mass *per* ion charge. Now since an independent method rendered the absolute value of the charge the masses of the ions are known.

Millikan's method yields most precise results because the small charges are bound to the drop which is orders of magnitude heavier than any atomic or molecular particle. Hence, the relative error of the mass determination is small.

What was given above was a simplified description of the experiment. In the real measurements the drop does not float but falls or elevates between the plates of the condenser exposed to the effect of friction. The velocity of the drop is measured, its mass being evaluated from that quantity.

As we saw earlier, Faraday did not accept the atom model of the chemists. Not even after his own results, which prove the atomic structure of both matter and electricity. Or, to be more precise, his results show that if one of them is of a discontinuous structure, so must be the other one as well.

Investigating the process of electrolysis Faraday observed that, whatever the substance was, the same amount of charge was always needed to separate a mass which equalled the quotient of its molar mass and its valence, M/z. The simplest of all electrolyzers are depicted in Fig. 12. The clock and ammeter refer to the fact that in order to know the amount of charge both current and time must be measured, $Q = It$. Faraday's law quoted above refers the amount of charge.

Applying the law to the example of silver and iron salt solutions it is stated that the same amount of charge is consumed for the deposition of one atomic mass (107.88 g) of silver from $AgSO_4$ and for the deposition of 1/3 atomic mass (55.85/3 g $= 18.62$ g) from $FeCl_3$. This amount of charge being called one Faraday is usually denoted by F according to precise measurements, its value is $1 F = 96,486 C$.

The atomic structure of both charge and matter offers an easy understanding of the law. The number of the elementary positive charges carried by a metal atom is equal with the atom's valence. This amount of charge must be neutralized for each atom at the negative electrode, called the cathode. Thus, F equals as many elementary charges as many atoms make one mole,

$$F = N_A e \qquad (2.9)$$

Since F is yielded by electrolysis experiments and e by the Millikan experiment, the Avogadro constant can be evaluated with high precision.

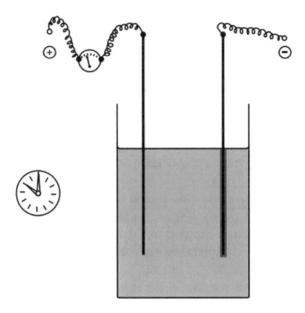

Fig. 12 Electrolyzer; metal is deposited on the negative electrode from the dissolved salt under the action of direct current

At present, we think this is the only correct way to understand the law of electrolysis. When everything is known that was written above were we not sure about the atomic structure of matter, if for example mass spectrometric results were unknown, we would have been unable to reach this conclusion. Moreover, had the Millikan experiment not convinced us about the existence of a universal elementary charge we could easily think that the charge *per* ion, F/N_A, is just an average value. Nothing would compel us to accept that each individual ion has the same charge. The existence of a non-divisible elementary charge could have been proven by an independent experiment only.

3   The Ideal of a Gas

Measurements were not too exact in earlier times. A lucky circumstance? Did this help the simple laws to be realized more easily because systematic thinking was not distracted by a number of minute details of secondary importance? It would be fruitless to muse on the extent of the inaccuracy and whether it is "useful" or "adverse"; the problem should meet the method of measurement.

What makes hydrogen and other gases similar? As far as chemical properties are concerned, they are certainly different; hydrogen burns, everything inflammable burns in oxygen, carbon dioxide extinguishes the flames. Chlorine makes hydrochloric acid with hydrogen, oxygen makes water. Talking about the difference of two gases, we usually refer to their chemical behaviour. But even their physical properties are often very different and some of the differences are very striking.

Their colour is an example, chlorine being pale green, bromine vapour dark brown. Their densities are also quite different according to Avogadro's law, densities being proportional to molar weights. That is, the volume of one mole gas, V_m is called mole volume and is equal with M/ϱ where ϱ denotes density. The volume, V of n moles of a gas is given as

$$V = nV_m \tag{3.1}$$

The gas volume is proportional to the amount of the substance, a statement which holds not only to gases but to any substance. Nothing more is stated here than the fact that if the amount of a substance is doubled, it takes up twice as much room which is not very surprising.

Experience shows that there are two further parameters which determine the volume of a gas: pressure, p and temperature, T measured on any arbitrarily chosen scale. These two state variables define the volume unequivocally thus one writes $V = V(p, T)$ where the parenthesis expresses that volume is a function of the variables.

That sort of functions is called state equation. Measurements show that this function is of the same form for each gas. The volume of each gas responds in the same way to a change in pressure or temperature. Gases are similar to each other in this respect. (That is a very imprecise statement, one should rather say that there exists a good number of gases the volumes of which behave similarly to the variation of pressure and temperature. Utmost exactitude, however, would be of little use to us at this stage.)

The results of the not too exact measurements can be expressed by the equation called perfect gas equation of state, as

$$V(T, p) = n\frac{RT}{p} \tag{3.2}$$

Here R is called the universal gas constant and T the absolute temperature. The gradation of T is centigrade with its zero point being below the melting point of ice 273.16 °C. The reason of this latter will be given soon. The unit of the absolute temperature scale is called Kelvin its symbol being K.

The equation of state duly expresses the Boyle-Mariotte law as well, showing that p and V are inversely proportional the proportionality constant depending on the temperature. The heat expansion can also be evaluated, understanding by that the relative volume increase upon one Kelvin increase of temperature if the pressure is held constant. If, for example, a gas is heated from 0 °C to T °C its volume increases from V_0 to V. According to the above equation of state the relative volume increase (it is relative to V_0) writes as

$$\frac{V - V_0}{V_0} = \frac{T - 273.16}{273.16} \tag{3.3}$$

That is, if the temperature differs from the melting point of water by 1 °C, the volume of the gas changes by its 1/273.16 part. This is the heat expansion of a perfect gas at 0 °C. The zero point of the Kelvin scale stems from here.

The variation of pressure with changing temperature at constant volume is given by a similar equation

$$\frac{p - p_0}{p_0} = \frac{T - 273.16}{273.16} \tag{3.4}$$

That is all that the perfect gas equation of state discloses about the behaviour of gases. Dictionaries call a state perfect or ideal when it can be approached but never attained. To be frank, no one has ever seen a gas which would behave exactly as stipulated by the equation. As a matter of fact, such a substance is beyond imagination. Let us consider the statements given in graphic form in Fig. 13. Imaginary experimental arrangements are also drawn below the graphs. They also hinted at the fact that here, as also in the following pages, pressure is meant to be hydrostatic, it being the same in every direction. The graphs show both V and p to decrease proportionally with decreasing absolute temperature finally becoming zero. Could one imagine a state where a body has no volume? That is against common sense. Let us perform our measurements with increased care, let us determine the heat expansions and compressibilities of a number of gases at different temperatures and under different pressures. The results become reassuring in the sense that do not imply any absurd consequences. But the simple form of the state equation valid for every gas gets lost. The exact measurements still offer some hope by showing that at small and smaller gas pressures the perfect gas equation of state proves to be a good and better approximation. The product pV is becoming proportional with T if p goes to zero the proportionality constant being the same for each gas, the product of the universal gas constant and the

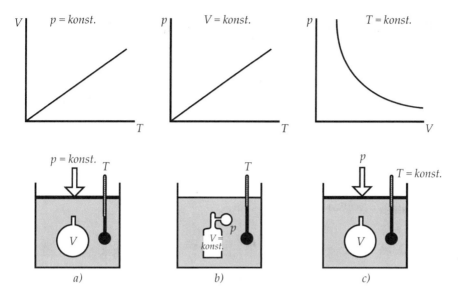

Fig. 13 Graphical representation of the perfect gas equation of state referring to the corresponding arrangement for measurement. **a** Varying V and T, p is constant; **b** varying p and T, V is constant; **c** varying p and V, T is constant

amount of substance,

$$\lim_{p \to 0} pV = nRT \tag{3.5}$$

The universal gas constant is the same for each gas. It is a quantity of physics with a physical content having a definite dimension. The word dimension does not refer now to the geometrical extension as in a sentence like "the flat disk is an object of two dimensions, the ball of three dimensions". The dimension of a physical quantity implies a method of determination. For example, *velocity* is said to be of the dimension *length/time* because it is found by measuring both a distance and the period of time which is needed to cover it, finally dividing the distance by the period. Or another example: the dimension of density equals *mass/volume* or *mass/(length)*^{3} because one obtains it by measuring the weight of the piece of a substance and dividing it by the piece's volume, which is proportional to the linear dimension of that piece: the radius of a sphere, the edge length of a cube etc. (As the end of the last sentence showed, the physical dimension can be regarded as a generalization of the geometrical dimension.) The equations of physics refer to measurable quantities with physical content, not to bare numbers. Therefore, the two sides of any physical equation express an equality not only in numbers but also in their dimensions.

The dimension of the universal gas constant, if multiplied by temperature, is equal to the dimension of pV, since n is a bare number. Hence, one may say that the dimension of R is *(volume · pressure)/temperature* with the unit litre·atmosphere/kelvin. Such units of measurement, and there have been several of them in the past, are frugal in revealing the physical content. Let us imagine a balloon being inflated from volume V_1 to V_2 with a gas of pressure p. The pressure is the same (hydrostatic) all over the surface, A hence force F equals pA. This force displaces the surface by a distance Δx performing work $L = pA\Delta x$ (let us remember, *work = force · displacement*). Considering Fig. 14, it becomes obvious that the product $A\Delta x$ is the same as the change of the balloon volume. Thus it writes,

$$L = pA\Delta x = p(V_2 - V_1) \qquad (3.6)$$

The product of pressure and volume has the dimension of work, pressure times volume change is the work done during expansion.

Although we wrote about gases in order to make the process easy to visualize, no properties of gases were made used of. The above relationship holds for any of the states.

Since the dimension of pV is work, more generally speaking energy, the dimension of R is *energy/(temperature · amount of substance)* its unit being J/(K mole) with its value 8.314 J/(K mole).

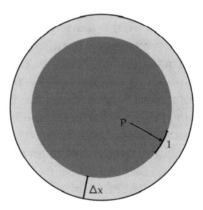

Fig. 14 The work of expansion is equal with the product of pressure and volume change

4   The Reality of Gases—From Gas to Liquid

The gaseous state as discussed above and given in Fig. 13 is an idealized limiting case for pressures approaching zero. The reality of the region of higher pressures is quite different, as can be seen in Fig. 15.

Volume is represented by the abscissa, pressure by the ordinate at different temperatures. Each curve refers to a constant temperature. The curves are called isotherms, this distorted Greek word refers to the constancy of temperature. The portion of the curve that describes the gas behaviour at high temperature is approximately a hyperbola and so are all the other curves at their low pressure regions. The hyperbola expresses inverse proportionality, thus these sections represent Boyle-Mariotte law as it was elaborated in the previous chapter. As p approaches zero this law holds better and better.

At low temperatures and high pressures, the curve deviates from the shape of a hyperbola and at even lower temperature it has three distinct portions: the hyperbola of the low p—large V portion goes over to a "horizontal" section which turns into a fast rising part. The "horizontal" section means obviously that the volume decreases without any pressure increase whereas the fast rise shows that only a very large pressure increase can bring about a small decrease in volume.

Let us consider the experimental observations, made by thermometer, manometer and also by the eye, the results of which the isotherms of Fig. 15

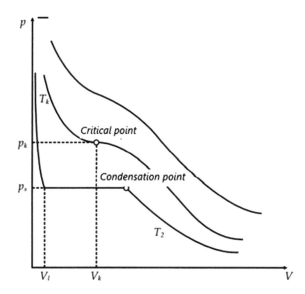

Fig. 15 The behaviour of gases in reality: pressure depending on volume at different temperatures

express. We should keep an eye on the gas! Let the gas be closed in a cylinder with a piston, as it is shown in Fig. 16, and let it be compressed at temperature T_2 by moving the piston down slowly. Nothing particular happens, V is seen to decrease as p increases, though the law of inverse proportionality, the function $pV =$ const. is not reproduced strictly.

Attaining pressure p_s (see Fig. 15) liquid drops form in the cylinder, the gas starts to be liquefied, the amount of the condensed substance is increasing at the bottom of the vessel. From that stage on the volume decreases without pressure increase (horizontal section in Fig. 15) because the response to decreasing volume is condensation and the liquid has a much smaller volume, i.e. a much higher density than the gas. Condensation proceeds at a constant pressure as the entire mass turns into a liquid. Liquids are known to be "almost incompressible", huge pressure is required for even a very small decrease of volume. Having attained volume V_l in Fig. 15 further volume decrease can be effected by very high pressures only; this is expressed by the steep part of the curve on the left hand side.

Now, it is important to know which gas is closed in the cylinder. The shapes of the curves differ for each gas, the temperature of condensation even more so. The figure tells us that at temperatures above T_k the horizontal portion which expresses condensation does not exist. Exactly at T_k the portion shrinks to a single point; the tangent at this point lies parallel to axis V and crosses the curve. With some expertise in mathematics one can happily notice the point of inflexion in this description. Indeed, there does exist a certain temperature where the curve $p(V)$ has a point of inflexion. At this temperature and pressure it is simply impossible to tell whether the substance in the cylinder is gas or liquid. An optical test is telling. If the substance is illuminated under such conditions, it appears as if the beam of light penetrated dense fog although no liquid droplets float in the cylinder. This effect, called critical scattering, is characteristic of the state. If the temperature is decreased even very slightly the liquid phase appears, if it is increased the gaseous state

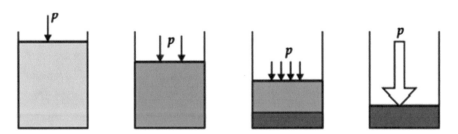

Fig. 16 Gases in reality: that is the way a gas behaves when compressed at a constant temperature

is established again. A substance at T_k and p_k is said to be in a critical state and the state variables are called critical temperature and critical pressure.

No gas can be liquefied above its critical temperature. Both T_k and p_k are different for each substance. The pair of critical variables of state are 33.3 K and 1.3 MPa for hydrogen, 154.4 K and 5 MPa for oxygen, 647.3 K and 22 MPa for water the scale being apparently wide. The critical temperatures of hydrogen, oxygen and a number of other gases are so low that it could not be attained in earlier times, thus these gases were thought to be impossible to liquefy having been called permanent gases. In view of the isotherms, this statement turns out to be naïve: every gas can be liquefied under its critical temperature.

It would be desirable to find a general equation of state valid for all and any real gas, a function $V(p, T)$ which, similarly to the $V = nRT/p$ for the ideal gas state, could describe the thermal properties of the gases at high pressures and also their behaviour as liquids. It is to be admitted that such a general equation of state is unknown although it is clear that every gas obeys similar rules of liquefaction, Fig. 15 being of general validity. The volume of each substance depends on temperature and pressure unequivocally a general function describing this dependence, however, is unknown. Despite a number of useful and resourceful approximations the solution is still not in our hands.

5   The Science of Possibilities—Thermodynamics

Measurements regarding atomic physics were not performed in the nineteenth century by the "atomic chemists" of the generation between Dalton and Kekulé, to call them by a name coined just now. There did not exist any appropriate method or instrument. Nevertheless, they were very ambitious in doing measurements, being the spiritual offspring of Boyle and Lavoisier. Their two most important instruments were the balance and the thermometer together with the manometer and the burette, one for measuring pressure the other for liquid volume. Such equipment enabled them to establish the notion of the perfect gas and this helped them to find the substances' compositions, volumes, and masses together with the thermal effects which go with the chemical transformations.

The notions of mass and volume were clear. Newton elucidated what inertial or gravitational mass means; the balance was known to measure weight, i.e. the product of mass and gravitational acceleration, the latter being constant at a certain spot and approximately all over the whole Earth. Thus,

it is all the same if one talks about mass or weight. Masses are additive properties: the mass of two bodies, measured either together or one by one is found
to be the same. Hence, atomists regarded the mass of a macroscopic body to
be the sum of the atomic masses.

With volumes things are more complicated. The macroscopic notion obviously takes its origin from our experience of space. It is, however, not obvious
how atomic volumes are added to the volume of a macroscopic body. The
question is whether atoms fill the space completely as the small coloured
cubes do in a Rubik-cube, or as grapes in a cluster with unoccupied spaces
between the neighbouring tiny balls, or are rather like flying specks of dust in
the air. Such questions cannot be answered directly by making use of balance
and burette only. Loschmidt's indirect answer took us to the estimation of
the Avogadro constant.

The interpretation of the thermometer reading is even less obvious. The
difference between heat and temperature has been established already by
Lavoisier and Laplace introducing the notions of heat capacity and specific
heat. If 10 g of hydrogen is burnt in an isolated steel vessel of a mass of
1000 g or 3000 g, or in an aluminium vessel of 1000 g, the temperatures
of the three vessels are different at the end of the process. Different vessels
are heated to different temperatures although the heat they absorb must be
the same. Since whatever the nature of heat is it is produced by a chemical
reaction, and if the reaction is repeated again and again evolved heat must be
always the same.

Such findings show that although heat causes a rise in temperature the
two are not identical. There are some other findings of importance. A piece
of ice melts in the vicinity of a hydrogen flame without any temperature
increase; heat might effect a change of state. A further relevant observation is
that if heat is furnished to a body its volume increases, the effect being called
thermal expansion. Also, heat can be transported from one body to another
and experience shows that it always goes from the warmer to the colder place;
that is, the warm site gets cooler and the cold site warmer. In this respect, heat
resembles a fluid; indeed, for quite some time heat was thought to be a fluid
and was called caloric, an idea shared and a name used also by great minds.

The trouble with this most pictorial concept is generally known. An
increase in temperature is not accompanied by any increase in weight consequently caloric must be a weightless fluid. It looks curious that the argument
used against phlogiston theory seems to have been forgotten until it was re-
invented to disprove caloric theory. Also, it was observed, and this was a more
decisive argument, that work can be transformed into heat. Rumford bore a
gun barrel with a blunt tipped drill which could not cause any change in the

material of either the barrel or the drill. Still the barrel got warm. Even latent heat can be produced by work as it was shown by *Davy* melting a piece of ice by friction. Thus it looked probable that heat is coupled with motion and work. If so, one can measure the amount of heat which can be produced to the expense of a certain amount of work.

Mechanical work equals the product of force and the displacement parallel to the force,

$$L = F(x_2 - x_1) \qquad (5.1)$$

If force F makes the body move from point x_1 to point x_2 work L is done. The amount of heat, Q is the product of heat capacity and of the change in temperature,

$$Q = C(T_2 - T_1) \qquad (5.2)$$

if the body of heat capacity C is heated from temperature T_1 to T_2. The remaining problem is the choice of units. Similar problem did not occur with the mechanical work since force is known to be the product of mass and acceleration. Before the physical nature of heat had been clarified both C and T had arbitrary definitions. The unit of temperature was subjected to the decision of one or another constructor of thermometers, *Fahrenheit*, *Réaumur* or *Celsius*. For heat capacity its unit was chosen equal to the heat capacity of 1 g water, probably because water was at unlimited avail and was easy to purify.

With these definitions at hand Q can be measured, heat and work can be compared. In his well-known experiments *Joule* stirred water and measured both the mechanical or electric work spent and the temperature rise caused. That way he determined the amount of heat produced by a certain amount of work. The result is of immense importance. Not as far as numerical value goes since it lacks any physical content in view of the arbitrary definition of Q. However, by this experiment it was proven that work can be transformed into heat completely. That is the reason why the nature of the body on which work is done plays no role; the amount of heat produced is the same if water is stirred or iron hammered. Heat was seen to be a physical quantity the same as mechanical or electric work. Joule was well aware of the theoretical importance of this finding. Being brought about by motion heat must be a kind of motion as well. The main result of his paper, containing a host of experimental details, was the determination of the quantity which he called the mechanical equivalent of heat. Still, the motto of his paper from the seventeenth century philosopher, John Locke was more far reaching: "*Heat is a very brisk agitation of the insensible parts of the object, which produces in us that sensation from whence we denominate the object hot; so what in our sensation is heat, in the object is nothing but motion.*" Joule's paper rendered the quantitative support of this idea on a macroscopic scale.

Somewhat earlier than Joule, *J. R. Mayer* came to the same conclusion by a medical observation and theoretical calculations. It seems almost incredible that he obtained the first impetus as a ship's doctor. He realized that healthy people's venous blood is of much lighter colour in the tropics than under temperate climate. As if blood were used off to a smaller extent when the air is warmer. As if metabolism used less work under conditions where it is easier to maintain the temperature of the body. The colour of venous blood might change by a number of other effects, so it is no wonder that the argument was not accepted with ease. However, by reading a later work of his it must be acknowledged without reservation that he was the first to formulate the law of energy conservation: "*Force, as the cause of motion, is indestructible.*" Later parts of the text make clear that the quantity he called force is the same as what is called energy nowadays.

Summarizing the above passages, experience tells us that, together with mechanical and electric energy, heat must be regarded as a kind of energy, thus energy conservation law must be extended also to heat. The mechanical energy of stirring is transformed into heat in Joule's experiment. A number of similar experiments can be performed with the invariable result that in a closed and isolated system: one with walls which are not permeable either to heat or to matter, the total amount of energy is constant. Denoting any kind of work by L, the above statement reads as $Q + L =$ constant. A change of the energy content from U_1 to U_2 is due to both heat and work,

$$U_2 - U_1 = Q + L \tag{5.3}$$

This equation, which expresses the conservation of energy while also taking heat into account, is the First Law of thermodynamics.

The reason of talking separately about heat is due to its particular properties. If mechanical, electric or any other kind of energy, with the sole exception of heat, is to be transferred from one system to another the energy must be converted into work first. If one wants to use the mechanical energy of a bent bow to chop an apple, the only way of doing that is by releasing the chord in order to accelerate the arrow. Work is done on the arrow which is used to chop as it hits its destination. The chemical energy of burning coal is converted into the mechanical work of a generator in order to get light energy from the lamp. The only exception is heat. This can be transferred from system to system in its original form, there being no need for the intervention of work. A warmer body can give heat directly to a cooler one if they are in thermal contact.

Heat is an exceptional form of energy also from another point of view. Any kind of energy can be converted into heat completely. Heat, however, can be converted only with limits into any other kind of energy. Let this be demonstrated through the simple example given in Fig. 17.

Let a gas, closed in a cylinder with a piston, be compressed from volume V_1 to V_2. The cylinder is in a large bath of constant temperature, called a thermostat, in order to maintain its temperature. The work done under pressure p equals $p(V_2 - V_1)$. During compression the piston rubs the cylinder wall and this inevitable friction creates an amount of heat, Q_1. The total amount of work during compression is

$$L_{compress} = p(V_2 - V_1) + Q_1 \tag{5.4}$$

Now let the gas be expanded. The work is the opposite of what was done during compression, $p(V_1 - V_2)$ whereas friction creates an amount of heat, Q_2.

$$L_{expand} = p(V_1 - V_2) + Q_2 \tag{5.5}$$

The second step made the system return to its initial state, still the work done during compression was not regained during expansion,

$$L = L_{compress} + L_{expand} = Q_1 + Q_2 \tag{5.6}$$

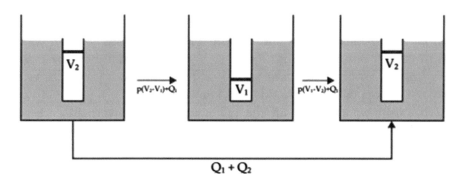

Fig. 17 The laws of thermodynamics are valid to any system and substance gases are taken as examples only to make the statements more visual

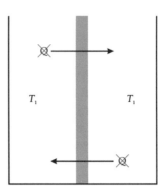

Fig. 18 No heat transfer without any temperature difference

Whatever the sign of the work is, positive when energy is added to the gas by compression or negative when energy is gained from it by expansion, part of the work is transformed into heat.

Let an experimental arrangement be as sophisticated as one can only imagine the result will always be much the same. Part of the energy turns into heat, which gets dissipated into the environment, and is lost for work. The contrary of that never happens; environmental heat, in the present case that of the thermostat, cannot produce work if thermostat and cylinder are at the same temperature. Sure, because heat flows only if there is a temperature difference (Fig. 18).

What is written above is the verbal version of the Second Law of thermodynamics. It seems as if it refers to the peculiarities of heat. On the contrary, it states one of the basic laws of nature, the law of irreversibility. Events, processes cannot be made undone.

That claim may sound striking, as if it contradicted our everyday experiences. Having climbed the hill, we can walk down; cold water can be boiled and cooled again; hydrogen and oxygen can be combined to form water, which can be electrolyzed into hydrogen and oxygen; water can be frozen into ice, ice can be melted into water—why to talk about irreversibility in nature if the opposite of each process is known?

Obviously, we can walk on the hill upwards and downwards, we can synthetize and decompose water, but not for nothing. More work is needed to decompose water than one can gain from the burning of hydrogen. An electric car consumes more electric energy uphill than it can obtain for loading the battery downhill. The difference is the heat spent by friction and dissipated into the environment. The cylinder with a piston discussed above is no textbook rarity but a simple model of this general theorem.

If so we have to consider the complete meaning of the word, irreversibility. It also implies that one of the directions is more favourable than the other. Let a system be left on its own, completely isolated from any outer influence, doing no work upon it, giving it neither heat nor substance. The First Law, that of energy conservation, would reveal much about the system. No more work can be done in it than the amount of heat which can be turned into work, or no more heat can be produced than the work done by any part of it. If the system is prone to any change of state a certain amount of substance might freeze or melt, increasing or decreasing the temperature, respectively. But the energy conservation law cannot predict whether heat is turned into work or work into heat, whether ice melts or water freezes.

Experience has taught us that if the temperature is the same in the whole system, heat cannot be turned into work, but work into heat can. Or, considering changes of state, if the system is warmer than 0 °C ice certainly melts, if colder water certainly freezes. Nobody has ever seen the contrary of that. Here we have the effects of a further governing law beyond energy conservation which selects the real process among those which are energetically possible.

It is heat which does this selection due to its particular nature. To make it short and, for the time being, somewhat sloppy, it is the amount of heat developed at the expense of all other kinds of energy that counts.

Let an example be proposed to this statement. In the spirit of this book let hydrogen be the protagonist. A simple apparatus is seen in Fig. 19.

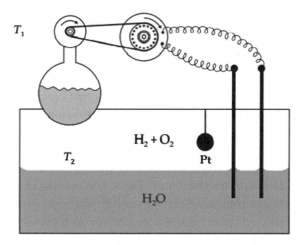

Fig. 19 Visualization of the Second Law. Water can be both formed and decomposed in this equipment. Work turning into heat and temperature getting uniform decide that water formation is the real process among the energetically possible ones

The closed and isolated box contains water, hydrogen and oxygen. There is a platinum sponge hanging in the gas space which is known to enhance water formation from the constituent gases even at room temperature. Water synthesis is accompanied by heat evolution, this heat can produce electricity via a steam turbine and a generator. Electricity decomposes water by making use of the two electrodes immersed. The process in the vapour phase, near the sponge, is this,

$$H_2 + \frac{1}{2}O_2 \xrightarrow{platinum} H_2O \tag{5.7}$$

In the liquid, near the electrodes, the electrolytic process is this,

$$H_2O \xrightarrow{electrolysis} H_2 + \frac{1}{2}O_2 \tag{5.8}$$

Water synthesis is followed by heat evolution which makes the system warmer. Water decomposition needs energy, so this process cools the water. That much can be learnt from the energy conservation law. But what about the direction of the whole process, does the amount of water decrease or increase? Let the water be warmer than the turbine at the beginning, if turbine works, electrolysis goes on. If the amount of water diminishes due to electrolysis the system gets cooler because the turbine obtains the necessary heat from the water and the gases. At the same time, the turbine gets warmer due to friction. Finally, electrolysis could go on if heat streamed from the cold water to the hot turbine; a process which contradicts to any previous experience expressed by the Second Law. The real process is the opposite: the catalytic combination of hydrogen and oxygen outdoes the electrolytic decomposition of water. This process continues until gases, water, platinum and turbine *plus* generator attain the same temperature. When temperature differences disappear no work can be produced from heat anymore.

Admittedly, heat production offers only rough information about the fate of the present system. The amount of heat evolved in the turbine depends on the construction and on the revolution of the engine; rate of heat evolution by the chemical reaction depends on the platinum surface and on gas pressure; heat transfer within the system depends on local temperature differences. Such eventualities influence the production of heat, rendering an exact physical definition impossible. In view of this, let a particular process be selected. It is the one which produces the least amount of heat, which should be divided by the absolute temperature of the system. This quotient is called entropy change.

If Q_r is the least possible amount of heat that can develop in a process taking place at absolute temperature, T, then the change of entropy as the system goes over from state 1 to state 2, $(S_2 - S_1)$ is given as

$$(S_2 - S_1) = \frac{Q_r}{T} \tag{5.9}$$

Even the least amount of heat is positive—whatever the details are, all kinds of energy turn into heat—hence $(S_2 - S_1)$ is always positive. The entropy of an isolated and closed system increases during any real process. The process gets halted when heat evolution stops, entropy does not increase anymore. The ending of entropy increase is not the cause but the sign of a process to be finished. Thermodynamic equilibrium is reached.

If the total energy of a system is constant the entropy change is either positive or zero,

$$(S_2 - S_1)_{U=\text{const.}} \geq 0 \tag{5.10}$$

That is the Second Law expressed in terms of the entropy function. The function, definition, properties and name were created in the mid-nineteenth century by *Rudolf Clausius*.

Under realistic experimental conditions, reaction vessels usually are not isolated, on the contrary, their temperature is maintained constant creating close contact with the constant temperature environment, a reliable thermostat. Hence, the energy of the reacting mixture varies because there is a continuous heat exchange between reactants and environment. The least amount of heat, Q_r which develops in the process can be expressed by the change of entropy as $Q_r = T(S_2 - S_1)$, and the First Law writes as

$$U_2 - U_1 = L + T(S_2 - S_1) \tag{5.11}$$

The energy change of the system is given by two terms. The first one is work the second heat which is inevitably given to the environment. The equation admits the fact that whatever is being done part of the energy is converted to heat.

Let us stop here for a moment in order to prevent some misunderstanding. The laws of thermodynamics have not been deduced on the previous pages. Some experiences were summarized and mathematical expressions were suggested to describe them. The First and Second Laws can be only realized. The word deduction would imply the existence of some more general laws the consequences of which are the laws of thermodynamics. Such more general laws are not known. Energy conservation and entropy increase are independent laws of nature.

There are few natural laws which play a more important role in the daily life of mankind than the two of these. Really, they are constant subjects of the press. The political effects of oil price, the race for hydrocarbon and uranium sources, local revolutions and international armed conflicts are all the consequences of our inability to produce energy. Anyone who needs energy must find its source. Energy can be transported, stored, transformed, used—but it cannot be produced. And with all the arguments among engineers, economists and environmentalist about the prudent use of the sources everyone knows clearly that it is only a fraction of the energy found in nature that can be made use of for practical purposes. Energy cannot be converted into work without any waste—part of the energy is dissipated as heat and all processes of the energy industry bring about contaminants. The question is only the scale.

World is held in the bondage of Joule and Clausius.

The world of course, not necessarily the Earth. There is a constant temperature difference between the surfaces of Sun and Earth which secures a continuous and practically inexhaustible flux of energy to our environment. All of terrestrial energy comes from the Sun with the sole exception of nuclear energy. Solar energy can be transformed into other kinds of energy, work can be done at its expense. All these finally turn into heat, which gets lost for useful work as it is demanded by the Second Law. But this waste is compensated by the continuous stream of Sun's energy transformed by green plants and nowadays by equipment which harness solar radiation.

One more thing. Those with an interest in the subject are often deterred from the notion of entropy due to two facts. On the one hand, its definition looks arbitrary; on the other, its content is far from pictorial. As far as the definition is concerned, let it be considered that thermal effects are controlled both by heat and temperature so it is small wonder that they appear here together. As a matter of fact, one could give many other definitions for an entropy-type quantity. The main point is that irreversibility, with the decisive role of heat, must be expressed and this can be done by a monotonous function which either increases or decreases.

Entropy changes in one direction in the course of all real processes similar to time. Some physicists maintain that whereas time is measured by stellar motion, that is by mechanical processes, since ancient times the direction of elapsing time is defined through the increase of entropy.

As to being pictorial it is true that nobody has ever seen entropy and no instrument for measuring entropy has ever been constructed. This is similar to energy which is neither visible nor directly measurable. One can measure force and displacement, current, voltage and time; thus, the change of one or another kind of energy can be determined. It is similar to the measurement of heat and temperature for the determination of entropy change in a certain process.

It is tempting to say that the difference between the notions energy and entropy stems only from daily usage since we call firewood an energy carrier and the bossy wife an energetic lady. This answer is, however, not completely honest. Better to say, it is correct to think of energy and entropy as notions of equal rank until the only sources of information are balance, thermometer and burette. From the perspective of macroscopic observations, both of them are used to express a basic law of nature, neither of them go back to further, perhaps simpler concepts. Later we will try to understand the macroscopic laws at an atomic-molecular level and then we will find important differences.

Until then let us finish this chapter with the words of Clausius, father of entropy:

The energy of the universe is constant.

The entropy of the universe tends to a maximum.

Provided the universe is a closed system tending to an equilibrium.

6   Pressure and Temperature—Inside

The branch of science whose basic concepts were set forth in the previous chapter is called phenomenological thermodynamics. This name refers to the fact that it deals with the laws of heat, work and temperature without trying to disclose the causes of the phenomena. Nowadays, this is regarded as the limit of the treatment. In the times when it was created it was called proudly the absolute theory of heat since its laws were formulated independently of notions which would come from different areas of physics. A number of excellent researchers were of the opinion that it would be a mistake to mar this beautiful edifice of intellect by any further efforts of explanation.

Other people, however, were of the opinion that the new basic principles expressed in the two laws are much too numerous and tried to reduce the results of thermodynamics to earlier knowledge. The best-understood branch of classical physics was mechanics, so an attempt was made to understand thermodynamics on that basis. Atomic theory was hoped to be a proper opportunity for that. The kinetic laws of atomic and molecular motion were

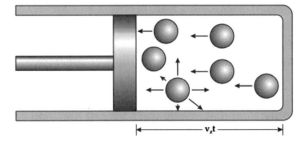

Fig. 20 Gas particles hit the walls of the vessel this being experienced as pressure

used to explain the underlying causes of all that was experienced by making use of balance, thermometer and burette.

The endeavour to find a molecular kinetic explanation for the thermody-namic effects had two general motives. One of them was the scientist's desire to explain observations through as few principles as possible. The other is the chemist's conviction that a chemical phenomenon can be claimed to be understood if its physical causes have been revealed. Physical chemistry was born from this conviction.

Heat and work were seen to be quantities of the same kind, i.e. of the same physical dimension. This is certainly true for macroscopic bodies. Let us try to translate this idea to the realm of microscopic particles. The first questions are whether pressure and temperature can be connected to the energies of atoms and molecules. Gases turned out to be appropriate substances for the demonstration of macroscopic laws, although it should be stressed repeatedly that these laws are valid for any substance in any state not only for gases. First, let gas pressure be studied in terms of the forces exerted by the constituent atoms. This idea is almost obvious, pressure being nothing else than force which acts at unit surface,

$$\text{pressure} = \frac{\text{force}}{\text{surface area}} \quad (6.1)$$

Now, let the particles in the cylinder be considered as they are represented in Fig. 20. The particles hit the piston exerting some force thus trying to move it outward.

To keep the piston at rest an equal force must act upon it from opposite direction. Force is known to be defined as the change of momentum during unit time,

$$\text{force} = \frac{\text{change of momentum}}{\text{time}} \quad (6.2)$$

This is tantamount to the better known expression *force = mass · acceleration*, because of the relationships *momentum = mass · velocity* and *acceleration = (change of velocity)/time*.

Let us examine an atom which travels towards the piston, hits it and then is repelled through an elastic collision. The magnitude of the velocity does not change but its sign does: velocity before the collision is $+v_z$ after it is $-v_z$. Thus, the momentum changes from $+mv_z$ to $-mv_z$ the momentum change being equal to $2mv_z$. Now we have to find the period of time during which this change takes place. A particle of velocity v_z travels a distance $v_z t$ during time t. Every atom being nearer to the piston than $v_z t$ hits the piston during a period which is not longer than t. To put this in a different way we may say that every atom within a box of the volume $v_z t A$ hits the piston until t, A denoting the surface area of the piston. If the number of particles in unit volume is ϱ this imaginary box contains $\varrho v_z t A$ particles, each of which undergoes the same change of momentum during period t, hence

$$\text{total change of momentum} = 2mv_z \varrho v_z t A \qquad (6.3)$$

As said above, pressure is given as *(change of momentum)/ (time · area)* thus pressure is calculated as

$$p = \frac{2mv_z \varrho v_z t A}{t A} = 2m\varrho v_z^2 \qquad (6.4)$$

It is, of course highly, improbable that all the atoms have the same velocity and move in the same direction. Instead of v_z we have to take an average velocity and also consider that given the six possible directions of motion it is only roughly the sixth of the atoms which move towards the piston. Summarizing all this pressure reads as

$$p = \varrho m \frac{\overline{v^2}}{3} \qquad (6.5)$$

where the bar above the symbol denotes average. Finally, by recalling that particle density means number of particles *per* volume, $\varrho = N/V$ we find the final result,

$$pV = \frac{2}{3} N \frac{\overline{mv^2}}{2} = \frac{2}{3} N \overline{\varepsilon} \qquad (6.6)$$

The kinetic energy of a particle being $mv^2/2$, symbol $\overline{\varepsilon}$ denotes the average kinetic energy of a particle.

The perfect gas equation has already been discussed, the left hand side of which was pV whereas its right hand side …

$$pV = nRT \tag{6.7}$$

N denotes the number of molecules, n the number of moles the two quantities being related to each other by the Avogadro constant, $N_A n = N$. Whenever the left hand sides of two expressions are equal so are also the right hand sides, thus we are compelled to accept that temperature and average kinetic energy of the particles are interrelated,

$$\frac{3}{2} n N_A \bar{\varepsilon} = nRT \tag{6.8}$$

or introducing a new quantity, called the Boltzmann constant $k_B = R/N_A$ the average particle energy is found as,

$$\bar{\varepsilon} = \frac{3}{2} kT \tag{6.9}$$

The task outlined at the beginning of the chapter seems to be completed. Having assumed that pressure is brought about by particles which collide with the walls of the vessel we could prove that the product of pressure and volume is proportional to the average kinetic energy of the particles multiplied by their number. This has taken us to the conclusion that the average kinetic energy is simply related to the temperature. The aim has been reached. Only one assumption was needed, it was the idea that pressure stems from atomic motion, the energy-temperature relationship appeared as a consequence of that.

That is an important achievement by which, however, the atomic basis of macroscopic thermodynamics has not yet been established. What has been understood is just the perfect gas equation of state in terms of molecular motion. At the outset a picture was devised for the atomic structure of a perfect gas. The gas particles were thought to be tiny, almost point-like spheres whose density is very low so they never hit each other, only the walls of the vessel. They neither repel nor attract each other otherwise their motion would be influenced by their mutual distances, something which was not taken into account. That was the model of the perfect gas which led us to its equation of state. It was pretty much what we made in one breath and that makes the basic idea less convincing. Less would have been more convincing. Either the material model or the energy-temperature relationship ought to have had independent support.

Nevertheless, the viability of a theory can be judged beyond the rigidity of formal logic. Model and calculation are thought to be robust if they can be applied to phenomena beyond the original problem, or even if they are able to presage earlier unknown effects. A good theory is predictive.

7   Predictions on Molar Heat

The perfect gas model as described in the previous chapter seems to be correct as far as the equation of state is concerned. The question arises whether it enables one to infer some further properties of the perfect gas. For example, its molar heat. This quantity is the heat capacity of one mole of gas: it is the amount of heat which increases the gas temperature by 1 K if no mechanical work is done; that is, if the gas volume is kept constant. (Let it be recalled: one mole is the mass which equals the molar weight expressed in grams).

According to our model, the energy of one mole of perfect gas equals the sum of the energy sum of N_A molecules,

$$U = N_A \bar{\varepsilon} = \frac{3}{2} N_A k_B T \qquad (7.1)$$

U changes only because T does, the two are proportional, hence molar heat can be given by a division,

$$C_v = \frac{U_2 - U_1}{T_2 - T_1} = \frac{3}{2} N_A k_B = \frac{3}{2} R \qquad (7.2)$$

Instead of heat, we thought of energy because no work was done, hence heat is the sole source of energy change. The subscript v refers to volume constancy.

The result requires some experimental check. As discussed above, gases at low pressure and high temperature obey the perfect gas equation of state (well, it is a very good approximation). Let us take gaseous helium as an example; let us measure its pressure and volume at different temperatures; thus, we can determine R. On the other hand we can measure C_v by determining heat input and temperature rise. Experience appears to be in splendid agreement with theory, constant volume molar heat is exactly one and a half times the universal gas constant. Moreover, perhaps more important than the sheer figures, molar heat does not depend on temperature, neither in experiment, nor in theory.

Fig. 21 Dumbbell model of a two-atomic molecule

The experiment can be repeated with other rare gases, argon, neon or krypton, the agreement is still reassuring. Both perfect gas model and the atomic understanding of pressure and temperature are strongly supported by these observations.

Now let us repeat the experiment for hydrogen, which is known to be a perfect gas at its best as to its equation of state. In spite of that the result is poor, molar heat being found to be much higher than $\frac{3}{2}R$. This may be sobering. Perhaps the entire idea of material properties to be governed by the motion of atoms, i.e. kinetic theory, is an error and all its consequences spring from sheer luck. That would be most unpleasant, fortunately this is not too probable. Or perhaps the molecular model of the perfect gas is mistakenly devised, or, finally, the model of elastic spheres is correct for rare gases but not completely so for hydrogen.

An obvious chemical fact might orient us here. Both classical molecular mass measurements and modern methods, for example, mass spectrometry, show that whereas rare gases are composed of atoms, hydrogen is of two-atomic molecules, H_2. If an argon atom is taken as a sphere, a hydrogen molecule might well be regarded to be a structure made of two spheres, similar to a dumbbell, as it is seen in Fig. 21.

Let the basic assumptions of the perfect gas model be upheld: the particles have only kinetic energy, neither attraction nor repulsion acting between them, their energies are independent of their mutual distances or positions, i.e. they have no potential energy. Helium atoms and hydrogen atoms move completely free, all their energies coming from this free motion. However, whereas a point-like atom can fly only in three different orientations, i.e. performs translational motion, a dumbbell can also turn around, rotate around its two free axes.

The average energy of translational motion in three dimensions is $\frac{3}{2}k_BT$. After some long observation one finds the three orientations to be equivalent: the particles move with the same average energy and keep their average orientation in each direction. Hence, each direction has the same share of energy, $\frac{1}{2}k_BT$.

Rotational energy is similar in this respect. Colliding molecules exchange their energies freely, they either fly or rotate; neither mode and orientation

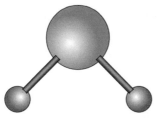

Fig. 22 Scheme of a three-atomic molecule, e.g. H_2S

of motion is more probable than the other. That is the reason why rotation around each of the free axes has the same share.

In short, energy is equally distributed among all kinds and orientations of motion, that is called the law of equipartition.

The average energy of a hydrogen molecule is higher than that of a helium atom due to its rotation,

$$\bar{\varepsilon}(H_2) = \bar{\varepsilon}(\text{translation}) + \bar{\varepsilon}(\text{rotation}) = \frac{3}{2}k_BT + 2\frac{1}{2}k_BT = \frac{5}{2}k_BT$$
(7.3)

The result can be compared with molar heat measurements. In line with what was written above the molar heat of hydrogen should be

$$C_v(H_2) = \frac{5}{2}R$$
(7.4)

and this was found indeed, within the error limits of the experiments. Theory reflects the temperature independence of C_v in this case too. Our basic concepts appear to be correct; nevertheless, the model of the substance had to be modified somewhat for the description of molecules built of two atoms.

A number of gases of two-atomic molecules were investigated and the molar heats of their majority was found to be $\frac{4}{2}R$. Let us not be discouraged by few exceptions for the time being, let us rather try to apply all that was used for atoms and two-atomic molecules to larger corpuscles: molecular kinetics and equipartition. Let us, for example, take hydrogen sulphide, the composition of which being H_2S and its structure depicted in Fig. 22.

As it is to be seen there are three free axes of rotation thus, different from what was stated about the H_2 molecule, H_2S can rotate around three different axes and, in terms of the law of equipartition, has a higher energy of motion by $\frac{1}{2}k_BT$.

$$\bar{\varepsilon}(H_2S) = \bar{\varepsilon}(\text{translation}) + \bar{\varepsilon}(\text{rotation}) = \frac{3}{2}k_BT + 3\frac{1}{2}k_BT$$
(7.5)

This is proven again through molar heat determination, the result of which turned out to be reassuringly as

$$C_v = \frac{6}{2}R \qquad (7.6)$$

The strength of the kinetic theory has been demonstrated, the molar heats of atoms, two-atomic and three-atomic molecules are correctly predicted both as far as figures and temperature independence are concerned, irrespective of molecular weights and chemical compositions. Moreover, the law of equipartition is shown to be valid. Provided the substance behaves as a perfect gas.

Is complacency now in order? Despite a few exceptions, can one state that molecules of a perfect gas move completely in the same way as macroscopic bodies do? Frankly spoken, molar heat calculations, splendid achievements as they are to support the kinetic theory of gases, are certainly not the last words on the subject. The not too rare deviations from the experimental data, as well as some observations that differ from theoretical expectation mark the limits of the theory.

Let us consider, just as one example, chlorine gas, which consists of two-atomic molecules, similarly to H_2. However, its molar heat is not equal to $\frac{5}{2}R$ and, what seems to be a more serious problem, it depends on temperature. This finding might raise suspicion regarding hydrogen and oxygen as well; indeed, their molar heats, measured in a wide temperature interval, also turn out to be temperature dependent meaning that average kinetic energy is not proportional to absolute temperature. Nevertheless, these gases are perfect at their best.

Is there any remedy to this malaise by any refinement of the molecular model? Perhaps the rigidity of the model must be eased and the rigid dumbbell replaced by two vibrating spheres, as if the atoms were not connected with a rod but with a spring, as it is given in Fig. 23.

The energy of such a vibrating system is higher than that of its corresponding rigid counterpart. Oscillation adds two further sources of energy beyond translation and rotation. The energy of a linear oscillator, that being the name of this structure, depends on how fast the masses oscillate and on

Fig. 23 The oscillator model of a two-atomic molecule

how near they are to each other or, expressed in a more professional manner, the kinetic energy of the oscillator depends on its velocity coordinate and the potential energy of the space coordinate. The law of equipartition also holds for oscillation, the same amount of average energy being allotted to each coordinate. The total average energy of a model molecule in Fig. 23 reads as

$$\bar{\varepsilon} = (3 \ + \ 2 \ + \ 2)\frac{1}{2}k_B T$$

[translation, 3 directions]   [rotation, two axes]   [oscillation, velocity and distance]

(7.7)

Thus this model predicts the molar heat to be

$$C_v = \frac{7}{2} R \qquad (7.8)$$

The molar heats of a good number of two-atomic gases at high temperatures were found in accordance with this calculation. Nevertheless, the above refinement still did not remove an important difficulty: it still predicts molar heat to be independent of temperature whereas experiments show the opposite. This independence cannot be removed from the theory until the law of equipartition is thought to be valid, that is the average energy of a molecule is given as $\bar{\varepsilon} = f\frac{1}{2}k_B T$ with f being the number of energy kinds that add up the total energy.

8   New Physics?

The molar heats of substances which otherwise behave as perfect gases depend on temperature indicating that under certain conditions the law of equipartition loses its validity. Previously, this law was introduced as a somewhat arbitrary conjecture and no exact proof will be given here either, still we try to show its fundamental nature.

First, let the perfect gas model be somewhat refined. This is inevitable because an important contradiction has been tacitly glossed over. On the one hand, the point-like particles were supposed not to hit each other, only the vessel walls. On the other hand, their average energy was stated to be proportional to temperature. Now, let us consider mixing warmer and colder portions of hydrogen gas. The natural observation is that the temperatures become balanced, producing a common medium level, which shows that the two portions have the same average energies. The process that takes the mixture to that state amounts to the acceleration of the slower and the slowing down of the faster particles. Energy exchange, however, must involve particle interactions; particles must collide.

Thus, according to a more correct picture, particles of a perfect gas must perform their motion mostly freely, paying no heed to each other, but, although rarely, sometimes they meet and on such an occasion they exchange their energies.

Energy exchange refers not only to the amount but also to the kind of energies. The particles loose and gain energies, translation might be changed into rotation, oscillation into rotation, rotation again into translation and so on. Equipartition means that the different kinds contribute equally to the total average, energy exchange is the same in each kind, the only thing that counts is the amount of energy at the moment of encounter.

If equipartition turns out to be invalid that points to the fact that the variation of different kinds of energies is not completely unlimited. It is as if it counted what kind of energy was being exchanged. It looks as if the probability of energy exchange were influenced by the states which exist not only before but also after the encounter. This is a curious idea that is worth wondering. Let us think of a car travelling uphill heading towards the valley behind the peak. If there is enough gas in the tank it can reach the peak and then it can roll downwards. In case of gas shortage, it gets stuck under the peak. The car's fate is decided by its momentary condition, not by any future one. What was said about the violation of equipartition would mean for a car that it could reach the peak with more or less difficulty depending on the depth of the distant valley. Cars are not like that.

Oscillating molecules are. Both calculations and measurements indicate that molar heats depend on temperatures mostly if also molecular oscillations are involved. Oscillations do not obey under any conditions, better to say at any temperature, the law of equipartition. Oscillating molecules often get or loose energies in a manner as if the oscillator knew its own future. Energy exchange is determined not only by the initial but also by the final states.

All our previous expectations based on cars travelling in the mountains, billiard balls rolling on the board, water flowing out of the glass contradict this statement. The car is ignorant about the valley behind the peak. Still calculations indicate that atomic oscillators, defying all what one would naïvely foretell, behave in a way very much different to what we are accustomed to in the world of macroscopic objects. The temperature dependence of the molar heats of gases can be understood only if it is assumed that energy exchanges of oscillators obey fundamentally different and unexpected laws.

Max Planck became a giant of twentieth century physics because he was the first to realize the difference between atomic and macroscopic oscillators. Later it turned out that what he saw was only the tip of the iceberg. With the eyesight of a genius.

9   Things Lost, Conserved and Born

One had better start with Newton's results in optics. Natural white light, like sunlight, white-hot iron or candlelight, was realized to show the colours of a rainbow once it passes through a glass prism. Glowing bodies were seen to radiate a mixture of lights. There ensued a centuries-long debate on what light is. It is beyond our present task to recount even the cardinal turning points of this development. Newton's original idea on particles of light was later superseded by the theory of electromagnetic waves due to *Young, Fresnel* and *Maxwell*. The velocity of wave propagation in vacuum was shown to be independent of frequency different frequencies corresponding to different colours. Finally, velocity of light in vacuum was shown to be the maximum which can be attained in nature, an idea upon which *Einstein*'s special theory of relativity is based. The only finding, important for our present consideration, is that light is nothing but energy that is emanated by some glowing body and propagates as a wave.

A physicist and a chemist, *Kirchhoff* and *Bunsen* employed the prism for more mundane tasks. They realized that whereas all glowing bodies emanate a number of different colours each substance emits a different selection. For example, the light rays coming from glowing iron or platinum are decomposed into different components by the prism. Sometimes this is visible even to naked eye: it is easy to distinguish the yellow light of the table salt, when thrown into the Bunsen burner, from the pale-violet light of potassium chloride or the red light of lithium sulphate. Thus, the components of the emitted light, the spectrum as it is called, reveal the material composition of the light source.

Having finished with their laboratory experiments, the two spectroscopists analyzed starlight; ever since the composition of the fixed stars are known as that of the Earth. (An anecdote of science history relates that after having completed the first analysis of the light of a star, Bunsen said: "*You, Kirchhoff, we have robbed the gods.*" He felt their discovery to be a Promethean feat.)

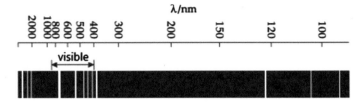

Fig. 24   Line spectrum of the H atom

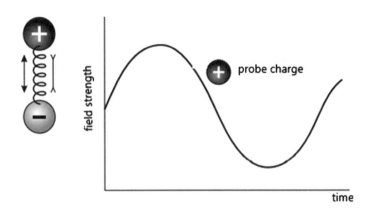

Fig. 25   That is how we try to imagine light propagation

The investigation of the spectra of low pressure gases was more than useful, it was surprising. Emitted light was seen to be decomposed into a number of distinct components, each of them with different frequencies, obviously. It was not the continuously changing palette of a rainbow with its colours melting into their neighbours that appeared behind the prism. Instead, a number of sharp light strips of different colours were seen. We stress the exceptional simplicity of hydrogen throughout this book. Figure 24 shows the spectrum of the emitted light of atomic hydrogen. It is a characteristic line spectrum; that is, the light is decomposed into well-defined, distinct components, nevertheless their number is quite high.

Now one could consider why the spectrum of the simplest substance is as involved as that, or why the components are of this or that colour (frequency) all these being clever and justifiable questions. The most important question remains: what is the reason of the spectrum to be composed of distinct lines?

Light energy propagates as an electromagnetic wave, the origin of which is the oscillation of the particles in the light source. This is symbolized in Fig. 25. The source is imagined as if small dumbbells oscillated in the source, similar to those used for modelling a two-atom molecule. The difference is that one of the balls is of positive, the other of negative electric charge. Such a structure is called an electric dipole.

Force is exerted by the two charges upon any electrically charged body or, to put it in a different way, an electric field of a definite field strength is present in the neighbourhood of the dipole. (The force acting upon a unit charge is called field strength.) If the dipole oscillates so does the field strength in its neighbourhood as well. The electric field changes everywhere in the space it takes, however, some time until the oscillation reaches a distant point; the probe charge in Fig. 25 follows the dipole oscillation but with a time lag. Sure, light does not propagate infinitely fast. The field strength travels with a speed of 3×10^8 m/s in vacuum independent of the dipole frequency.

The energy of light can be measured by two methods as they are proposed here. One of them is rather straightforward. A light-absorbing material, e.g. a platinum plate covered with platinum black, is put across the light beam and the temperature increase of the plate is determined after a certain period of time; the simple apparatus is seen on the left hand side of Fig. 26. Recalling what was said about calorimetry, the equivalence of heat and work, the total energy absorbed can be evaluated from the temperature increase if one knows the heat capacity of the plate. Let the total energy be denoted by U_Φ.

The other method is based on the fact that light can eject electrons from a metal (Fig. 26, right hand side). Electrons may leave a metal plate in vacuum if it is illuminated, a process called photoeffect. Ejected electrons can be collected in a positive electric field; thus, the electrons can be counted.

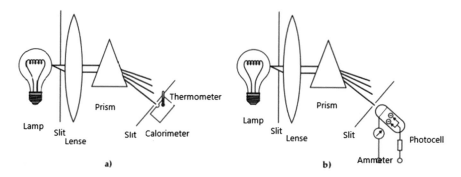

Fig. 26 Simple light experiments. **a** Light energy measured by calorimeter, **b** electron ejection from metal by light

The positive grid collects n electrons during illumination time, t. Each electron having charge $-e$ the total negative charge is $-ne$. Since electric current, measured by the ammeter, equals the charge divided by time, the current is given as

$$I = -\frac{ne}{t} \tag{9.1}$$

Since $-e$ is known by the Millikan experiment and t can be measured by our wrist watch (or some more refined time keeping instrument) the number of photo-ejected electrons can be determined.

Let the calorimetric light energy, U_Φ be compared with the number of electrons, n. Both of them are found to be different for each and any light of a different colour, that is of different frequency. A finding of fundamental importance is that the quotient, total energy to the number of electrons, is proportional to the light frequency,

$$\frac{U_\Phi}{n} = \hbar\omega \tag{9.2}$$

(Here, customarily, we used instead of frequency, ν the angular frequency, $\omega = 2\pi\nu$.) The proportionality constant, $\hbar$ is independent both of the illuminating light frequency and the nature of the illuminated metal. It is a basic constant of nature, like light velocity or electron charge. This constant, multiplied by 2π is called Planck constant, its symbol being $h = 2\pi\hbar$ and its value $h = 6.626 \times 10^{-34}$ J s.

Energy must be imparted to each electron to make it leave the metal. The above results show that energy per electron equals $\hbar\omega$, this being the least energy light can lose, a metal electron can absorb. In short, photoeffect shows that light energy is quantized. The indivisible part of light energy is the portion, $\hbar\omega$ called photon with a reminiscence of the name of the electron.

Having learnt all this, we must sincerely wonder at the line spectrum of the H atoms. Different frequencies correspond to different energies, energy and frequency being proportional to each other. Thus, H atoms can only emit certain well-defined amounts of energy, neither lower nor higher. As if a high jumper could pass the bar either at 1.20 or 1.90 but knocked the bar off at any height in between.

We shall be even more amazed if we recall what we learnt at secondary school about the simplest oscillator, the kind shown in Fig. 25. Let the angular frequency of the oscillator be denoted by ω, the aggregate mass of the oscillating bodies by m, and let the spring be expanded to z_0 at the start of the motion this initial elongation being called the amplitude of the oscillation. The total oscillator energy is given as

$$\varepsilon_{\text{osc}} = \frac{1}{2}m\omega^2 z_0^2 \tag{9.3}$$

The energy of a single photon, as understood through the photoelectric effect is

$$\varepsilon_{\text{photon}} = \hbar\omega \tag{9.4}$$

Since light is emitted by the oscillator the two energies are necessarily equal, $\varepsilon_{\text{osc}} = \varepsilon_{\text{photon}}$. It follows that frequency and amplitude are not independent,

$$m\omega z_0^2 = 2\hbar \tag{9.5}$$

That is something one would not expect! Let us take a spring, expand it to a certain length (amplitude) z_0, and let it be left on its own. It oscillates with a certain angular frequency, ω. Now let this simple observation be repeated with a number of different amplitudes, the spring is always seen to oscillate with same frequency. Amplitudes and frequencies of tangible, macroscopic springs are independent of each other. Were the H atom an oscillator of that kind it would emit every frequency, the complete rainbow would appear behind the prism, no line spectrum would ever be seen. But no, amplitudes and frequencies of the atomic oscillators do depend on each other as it is witnessed by the line spectrum or the photoeffect.

This mutual relationship can be expressed also through elongation, z and momentum, p. As it is known, *momentum = mass · velocity* that is $p = mv$, and the maximum speed of the oscillating body is $v_0 = z_0\omega$. Making use of these expressions the above equation can be written as

$$p_0 z_0 > \hbar. \tag{9.6}$$

The product of the highest momentum and highest elongation of an oscillator is always larger than a universal constant of nature. No such limitation has ever been observed in any macroscopic system, its momentum *and* elongation might be of any value, small or large, together.

We tried to arouse wonder in the previous pages by several occasions. Infringing equipartition indicates that a particle's fate is influenced by not only its present but also its future state. The line spectra of gases show that atomic oscillators radiate only several, well-defined energies. Photoelectric phenomena prove the atomic structure of light and that the smallest portion of energy of light energy is proportional to the frequency. These experiences, together with a number of further ones which were not discussed above, cannot be understood within the framework of classical physics the laws of which were established for macroscopic bodies.

The great discovery of Newton was that the moon and an apple obey the same laws. The equally great discovery of Planck, Heisenberg and Schrödinger was that an atom and an apple obey fundamentally different laws. It is difficult to tell which one of the two statements is more surprising.

New laws of physics must be established by considering the behaviour of the electrons or atomic particles. These laws have a twofold task. On the one hand, they have to express the experimental results in atomic physics. On the other hand, they must maintain the contact between microscopic and macroscopic objects, that is, between atomic particles and the instruments for observation. For that sake, notwithstanding the fundamental differences between old and new physics, conservation of energy and momentum, and the Coulomb law prevail.

"*The Book of Nature is written in the language of mathematics*"—that is *Galileo*'s great legacy to every scientist. A law of physics is a mathematical expression which refers to measurable quantities. Take, for example, the law of the two-armed lever. It tells us the conditions of equilibrium: the weights and lengths of levers at which balance is established. We all know the law since the time of *Archimedes*. Given the lengths of the arms to be l_1 and l_2 the ratio of masses M_1 and M_2 are given by the equation

$$\frac{M_1}{M_2} = \frac{l_2}{l_1} \tag{9.7}$$

This is an algebraic equation a type of which is sufficient to describe simple mechanical systems at rest.

Laws for moving bodies or, more generally speaking, for phenomena which change in space and time cannot be formulated exclusively by algebraic equations. For that a mathematical tool is needed that can assign the actual state of the system—its velocity, acceleration, energy, density, temperature or other qualities—to a point in space and moment of time.

The concept of the mathematical function with all the relevant operations are appropriate for that task. It is no wonder that mathematical analysis and calculus was created in the times of and in good part by Newton. Also, the definition of velocity was his achievement for cases when it changes every moment. Let a body move from point x_1 to x_2 during a period of time $(t_2 - t_1)$. Velocity, v is the quotient of displacement over time; if the quotient changes throughout the motion we have to make the period shorter and shorter in order to obtain a meaningful result. Mathematically speaking, velocity is defined as the limit of the quotient,

$$v = \lim_{(t_2-t_1)\to 0} \frac{x_2 - x_1}{t_2 - t_1} = \frac{dx}{dt} \tag{9.8}$$

Such limit types are called the first derivative or first differential quotient of a function. Velocity is defined here as the first derivative of displacement with respect to time.

Momentum, defined as a product (mass times velocity), reads as

$$p = m\frac{dx}{dt} \tag{9.9}$$

Force, F according to Newton is the first time-derivative of momentum,

$$F = \frac{dp}{dt} \tag{9.10}$$

Considering the definition of p, this equation takes us to the derivative of a derivative i.e. the second time-derivative of displacement,

$$F = \frac{d}{dt}\left(m\frac{dx}{dt}\right) = m\frac{d^2x}{dt^2} \tag{9.11}$$

writing it with the generally accepted mathematical symbols.

Functions have proved to be most appropriate for the processes of classical physics where quantities usually vary in a continuous manner. A dumbbell can rotate with any speed, an oscillator can oscillate with any frequency, any amount of energy can be absorbed by a substance. The physics on the atomic scale is different, some experimental facts having been presented in the previous pages, which show that certain physical properties of atomic systems vary in a discontinuous manner. Jumps are often observed, that is why this discipline is called quantum mechanics. Continuous functions cannot be used for the description of such phenomena, a new mathematical tool had to be applied for that sake. Operators can do the service.

In Newton's classical world physical quantities appear as functions of space and time, a function assigning number to number. Operators assign function to function. An operator is an instruction in mathematics like the sign of multiplication, of root extraction, $\sqrt{}$ or of derivation, $\frac{d}{dx}$. An operator, if applied to a function, renders another function. In quantum mechanics neither numbers nor functions but operators are the protagonists. This is perhaps not more surprising than the introduction of limits and derivation was 300 years ago to minds accustomed to algebra and geometry.

The conceptual difficulty lies with an operator being meaningless until it is not applied to a mathematical entity. A number must go under the root, derivation must be performed on a function etc. Physical quantities like e.g. momentum or energy are coupled with an operator, one for each, which is applied to a function.

The right mode of thinking, according to Newton's classical physics, is the following. A body is accelerated whenever it is exposed to the action of force, F. The relationship between force, mass and acceleration is given by the equation above, for the simplest case of a point-like mass. It refers to one spatial dimension for simplicity's sake alone. Force might vary both in time and in space, thus the left hand side is a function of x and t, $F = F(x, t)$, whereas the right hand side is a derivative. That is what a differential equation looks like. Methods for the solutions of differential equations are well known. The solution is not a number or set of numbers, as in the case of algebraic equations, but a function, in the present case it is the function $x(t)$. This tells the position of the body, x at moment t. Generally speaking, the solution gives the path of the body.

Quantum mechanics, as developed by Heisenberg and Schrödinger is very different from that as far as both methods and answers are concerned. Let $\hat{H}$ denote the energy operator and ψ the function upon which the operator acts. $\hat{H}$ consists of multiplications and derivations; this, however, is not our present concern. It should be stressed that it does not contain time as a variable, so no dynamical information, no time-dependent solution can be expected. The task is quite difficult, even with this simplification. Let the object to be discussed be an electron, a molecular oscillator, a hydrogen atom or any other microphysical structure, its energy, E is determined by the equation

$$\hat{H}\psi = E \cdot \psi \tag{9.12}$$

The product (energy times ψ) is equal to the result of the operation $\hat{H}$ applied to ψ. Here the energy operator $\hat{H}$ is given by the general theory and the structure of the object discussed whereas both E and ψ are to be found by solving the above equation.

This type of expression is called eigenvalue equation. Methods for the solutions of eigenvalue equations are well known. (The word *eigenvalue* is half-German and half-English as a relic of the mother tongue of the creators of quantum mechanics, *eigen* meaning *self*.) Usually the solution is a set of eigenvalues, E and eigenfunctions, ψ with one or more functions belonging to a given eigenvalue,

$$
\begin{aligned}
&E_1 \ \psi_{11}, \ \psi_{12} \cdots \\
&E_2 \ \psi_{21}, \ \psi_{22} \cdots \\
&E_3 \ \psi_{31}, \ \psi_{32} \cdots \\
&\ \vdots \quad \vdots \quad \ \vdots \quad \vdots
\end{aligned}
\tag{9.13}
$$

These numbers and functions disclose everything about the system within the possibilities of quantum mechanics. The quantities E are equal to the possible energy values of the system, whereas the absolute squares of the functions, $|\psi(r)|^2$ give the probabilities of the particle to reside in the neighbourhood of site r. The solution renders the possible energies and the probable domains of position. Not a single word about certainty.

Forces, like Coulomb attraction between charged particles or the elasticity of an oscillator determine both energy and eigenfunction through the energy operator. Lowest energies and most widely spread-out functions go together.

The variations of the states as time proceeds, that is the dynamics of a quantum mechanical system can be treated on similar lines; mathematics becomes more involved but the idea of operators, eigenvalues and eigenfunctions play similar roles. The character of the solutions is similar. No certain path of a particle, $r(t)$ can be obtained, only probable paths can be predicted.

A comparison of calculations and experiments, performed over the more than ninety years that have passed, convincingly proves that the above-sketched methods for the evaluation of energies and probabilities are in complete agreement with observations. Frequencies and intensities of spectra of gases, strengths of valence bonds, currents in a transistor can be evaluated, as well as the relationship between the energy and frequency of an oscillator, a problem mentioned at the beginning of this chapter. An explanation can be found for the violation of equipartition as well.

However, is quantum mechanics unable to describe the path of a particle, for example that of an electron in the photoelectric experiment (Fig. 26)? Surprising or not, no effort has ever been made in that direction. It can tell the probability of an electron to take one path out of many, but does not try to tell the position of a certain electron at a given moment.

As modern physics was taking shape, even the greatest of physicists were of the opinion that the tools used in classical theories can be made appropriate for the description of the microscopic world. They hoped that phenomena like for example specific heat anomalies, the existence of line spectra, or the quantized nature of light absorption might be explained within the framework of the notions force-path-time. All such endeavours failed. Even the most resourceful and boldest extensions of classical physics turned out to be insufficient for the correct description of the events on a microscopic scale.

Some basic laws of classical physics, however, turned out to be valid even in quantum mechanics. The conservation of momentum, angular momentum and energy are valid in the realm of microphysics as well. These conservation laws have never been questioned by any experiment and are basic constituents of every theory.

One had to abandon some notions of classical physics but that was no real loss. The task of a physical theory is to propose a mathematical framework for the observations. The notion of path was a natural part of Newton's physics since the path of a rolling stone or a flying grain of dust can be measured. The path of an electron is impossible to measure in that sense. The reason is, simply, that the very measurement makes both path and velocity change. For example, the position of an electron could be determined by illuminating the object and measuring the direction of the scattered light. Sure, but light gives energy to the electron influencing both its path and velocity. And light must act upon the electron, otherwise no scattering would take place.

One of the most spectacular and most important results of quantum mechanics is the prediction of the smallest uncertainty by which the position and velocity of a particle can be determined simultaneously. That limit is independent of any method of measurement. Let Δp_x denote the uncertainty of the momentum in direction x and Δx the uncertainty of the position; the inequality below always holds,

$$\Delta x \, \Delta p_x \geq \frac{\hbar}{2} \tag{9.14}$$

The more precisely a particle's position is known, the less certainly is its momentum known. If Δx is small, Δp_x is large, and also the other way round: a more exact determination of the momentum necessarily goes together with a less exact knowledge of the position. That is Heisenberg's uncertainty relationship. It is to be stressed that it is not the consequence of some inadequacy in the method or equipment of measurements. It expresses the nature of information one can obtain of microphysical objects. What was said about the position of an electron before, is expressed here in quantitative terms: the position can be precisely determined with small wavelength light in a scattering experiment, which corresponds to large photon energy, which again changes the electron velocity markedly.

Writing about photoelectric phenomena and atomic oscillators, we showed how experimental facts forced us to give a lower bound to the product (displacement times momentum). The uncertainty relationship, reflecting the conclusion of a huge number of experiments, obviously refers also to atomic oscillators.

Another relationship similar both in form and content to the above one refers to the energy and lifetime of a system's state. We repeat the syntax of a previous clause with different physical symbols: let ΔE denote the uncertainty of energy and Δt the uncertainty of time; the inequality below always holds,

$$\Delta E \, \Delta t \geq \frac{\hbar}{2} \tag{9.15}$$

The content is similar: the more precisely it is known how long a state prevails, the less precisely is known how large its energy is.

Quantum mechanics describes correctly everything that can be measured and stays silent on everything that cannot be observed.

10   Rule and Exception

Having investigated the molar heats of gases we had to realize it had been an error to believe that atoms or electrons behave as macroscopic bodies do. We were compelled to understand quantum mechanics, the "new physics." Still called new, almost after a century of their basic laws having been established? We were writing about specific heats in this context because that is the way a physical chemist thinks. Max Planck, the first great scientist of the field, however, was forced by the properties of light emitted from glowing bodies, i.e. of electromagnetic waves to realize that the oscillations in the realm of microphysics are very different from the world of macroscopic springs and chords.

Our present task is simply to consider, once again what was written about the particles of a perfect gas, with the difference that now the laws of quantum mechanics are to be applied. This must take us to the correct result, offering an explanation for the deviation of the molar heat from our earlier expectations. Knowing the laws of motion of each particle, describing their encounters and energy exchange processes one by one, we can tell the aggregate behaviour of the particles, the behaviour of the gas.

This program cannot be fulfilled. Not because the laws are incomplete but because the molecules are much too numerous. No human being and no supercomputer could find the solutions of so many equations, could register as many results as it is needed to describe the motion of 10^{23} particles. As a matter of fact, this is not our real task either, since we are interested in the common properties of the multitude of the particles, not in the individual fate of each.

If one has to get some insight into the nutrition of the population in a country one could ask each citizen every day about her menu. That would be tedious, tiresome and would also yield irrelevant information. It plays no role in the country's meat consumption whether the Millers had a roast-beef on Monday. Statisticians have a good command of methods which give information about average meat consumption per year without peeping into each housewife's pot.

Similar methods were developed by the physicists in order to understand the behaviour of macroscopic bodies on the grounds of the atomic laws of motion. The branch of science that studies multi-particle systems is called statistical physics. Here, one has to know all the laws which govern the motion of the individual particles and must forget everything that has no bearing on the common behaviour of the multitude. Let it be repeated, the relevant common properties of a multitude are those that can be observed by macroscopic methods, roughly speaking, the ones that can be measured by balance, thermometer and burette.

Could we single out a hydrogen molecule and examine it for a longer period of time we would realize that as moving, colliding with other molecules, hitting the wall of the vessel, rotating and oscillating, it gains and loses energy. Its energy content varies all the time, its velocity and total energy very rarely remain constant. We should not try to follow the energy changes of each molecule. Let us be content to know the way energy is shared among the molecules: let us count the molecules which have certain amounts of energy. Using a technical term, let the energy distribution of the molecules be determined.

What can we observe casting a quick glance on the multitude? Does the same amount of particles fall to each energy (Fig. 27a)? Or is the opposite true, has each particle the same energy (Fig. 27b)? Or, perhaps, the higher the energy is, the more particles falls to it (Fig. 27c)? Or a good many particles have low energies and only a few of them are of very high energy (Fig. 27d)? We can find a probable answer but no certain one. Only the most probable distribution of energy can be determined. The question is meaningful only if some limiting conditions are stipulated. To that end, both the number of particles and their average energy are taken as constant. Thus, the system

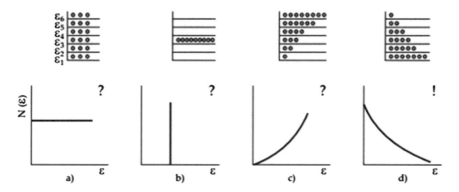

Fig. 27 About the energy distribution of a multitude. **a** Is the energy irrelevant? **b** Does each particle have the same energy? **c** Do few particles have low energies, and many of them high energies? **d** Do many particles have low energies, and few of them high energies? Yes, **d** is true! $N(\varepsilon)$ is the number of particles whose energy is ε

under investigation is in equilibrium, its macroscopic properties do not vary in time.

The perpetual energy exchange between the particles renders the cases of Fig. 27a, b) highly improbable. Indeed, it is not impossible that either all the particles have the same energy, or the same amount of particles falls to each and any energy at some given moment, but no doubt the first collision would mar either of these situations.

If high energy particles are plentiful and low energy ones are few it is difficult to imagine the average energy to remain constant. For example, when a high energy particle loses one-hundredth of its energy, a great number of low energy particles must increase their energies in a similar proportion at the same time in order to maintain the average. Hence, Fig. 27c might not be the right distribution either. One would expect rather the opposite.

One may have the feeling that the more particles share the less energy the collisions have an easy and easier task to keep the average energy constant as energy is perpetually exchanged. It is easy to keep a balance if all that happens are small changes of small energies. The most probable state seems to be represented by Fig. 27d: a high number of particles are of low energies with high energy particles being relatively rare.

Detailed calculations result in the exact shape of the "most probable" curve of the type sketched in Fig. 27d. Function $p(\varepsilon)$ tells the fraction of particles of energy ε. Out of N particles there are N_ε of energy ε, thus $p(\varepsilon)$ means

$$p(\varepsilon) = N_\varepsilon/N \qquad (10.1)$$

Calculations show $p(\varepsilon)$ to have the important property: when ε increases by a small amount, $\delta\varepsilon$, then p decreases by δp which is proportional to both p and $\delta\varepsilon$. This writes as

$$\delta p = -\frac{1}{k_B T} p(\varepsilon)\delta\varepsilon \tag{10.2}$$

Here k_B and T denote Boltzmann constant and absolute temperature, respectively, as before. This expression is quite graphic: the more particles have a certain energy the greater bearing the change of this energy has on the multitude.

If δp is much smaller than p and $\delta\varepsilon$ is much smaller than ε than the above expression can be regarded as a differential equation,

$$\frac{\delta p}{\delta\varepsilon} \cong \frac{dp}{d\varepsilon} = -\frac{1}{k_B T} p(\varepsilon) \tag{10.3}$$

Its solution requires a function the first derivative of which is the function itself multiplied by -1. This is a well-known function, mathematicians call it natural exponential function, its symbol being e^{-x}. One of its most important properties refers to its first derivative,

$$\frac{de^{-x}}{dx} = -e^{-x} \tag{10.4}$$

The quantity e is an infinite, non-recurring decimal being equal with 2.718....

By making use of this function the probability of occurrence of energy ε writes as

$$p(\varepsilon) = \frac{N_\varepsilon}{N} \propto e^{-\varepsilon/k_B T} \tag{10.5}$$

with the sign $\propto$ denoting proportionality.

The validity of this expression can be directly checked in some simple cases. Let us take air density as an example. Gravitation causes acceleration g acting upon a particle of mass m. Hence, the potential energy of the particle at a height z equals $\varepsilon = mgz$. According to the above equation, the number of particles in a fixed volume varies with height z as

$$N(z) = N_0 e^{-mgz/k_B T} \tag{10.6}$$

Here N_0 is the number of particles at height $z = 0$. We may know m from mass spectrometry (atomic particles) or by measuring particle diameters under a microscope (small droplets), g from gravitational pendulum experiments, k_B from the perfect gas law and the Avogadro constant, and T from thermometer reading whereas z can be measured with a yardstick or a theodolite. The number of particles in a given fixed volume is proportional to particle density. Thus, all the quantities that figure in the above expression can be measured by macroscopic methods, and the expression can be checked directly. Indeed, if temperature is held strictly constant the particle density varies according to this equation called barometric formula.

The law is valid with no respect to either the substance or the size of the particles. The density of grains of dust in the air changes in a matter of several tenths of a metre to a perceptible extent, whereas the molecules of the atmosphere show a measurable density change on hundred metres or so. Most detailed measurements were performed on grains of a resin floating in a liquid with the height scale of observations of some $100~\mu m$.

The barometric formula has been proven experimentally with high precision still air line pilots and mountaineers are dissuaded to use it. The air is cooler on the hilltop than at the lakeside, T is not constant.

The exponential function $p(\varepsilon)$ is called Boltzmann distribution in honour of the great Austrian physicist of the second part of the nineteenth century. It was he who created, in parallel with *Maxwell* and *Gibbs*, the concept of statistical thinking that enables physicists to make exact statements on systems that consist of a huge number of particles. Exact ones but not certain at the same time, the repeated use of the word probability is not by mere chance.

First of all, we have to understand the way and the content of how Boltzmann distribution characterizes the energy conditions of a system. Energy might be distributed among the molecules in many different ways (cf. Fig. 27), the question is what is the probability W of the prevalence of a given distribution. It was shown by rigorous mathematics that Boltzmann distribution is the one which prevails; its probability W_B being much higher than that of any other distribution, provided the system is in equilibrium. Let heat and material transport, chemical reaction and any other temporal process be excluded, in such cases it is most probable that the energies of the molecules obey Boltzmann distribution. The reliability of this statement depends on the size of the system: were it infinitely large, the probability of this equilibrium distribution would turn into certainty. Nevertheless, systems that are regarded as small as a grain of dust or a raindrop are made of a huge number of molecules so even in these cases Boltzmann distribution "can be taken for granted".

If the system is not in internal equilibrium, for example some regions are warmer, others cooler, then the actual energy distribution is different from Boltzmann's. The probability of any non-equilibrium distribution is always lower than the equilibrium one,

$$W_{\text{nonequilibrium}} < W_B \qquad (10.7)$$

Thermodynamic equilibrium is the state where all temperature differences are levelled out, no currents flow, neither work is being transformed to heat, nor heat to work, and compounds are neither formed nor decomposed. In equilibrium state the probability of the energy distribution equals W_B. As a system in a non-equilibrium state approaches equilibrium, W tends toward W_B.

This description has a curious semblance of the entropy function defined in macroscopic, phenomenological thermodynamics. Entropy increases as long as the system approaches equilibrium and is constant from that moment on. It seems to be evident that the probability of the energy distribution, W and entropy, S are closely related.

The relationship is easily revealed. Entropy being connected to heat evolved or absorbed is proportional to the amount of substance that makes up the system. A system twice as big develops twice the amount of heat provided all other conditions are kept unaltered. Dividing a system into two in thought, the entropies of the two parts, S_1 and S_2 must be added in order to obtain the entropy of the whole system, S,

$$S = S_1 + S_2 \qquad (10.8)$$

Probability calculus teaches that if the probability of an event in a subsystem is W_1, that in another subsystem it is W_2 and the events in the subsystems are independent of each other, then the probability that both events take place is equal to the product of the probabilities referring to the subsystems,

$$W = W_1 W_2 \qquad (10.9)$$

Let us consider, the probability of a dice thrown to show six is $\frac{1}{6}$, a penny tossed to show head is $\frac{1}{2}$. If both dice and penny are tossed the probability of finding six and head is $\frac{1}{6}\frac{1}{2} = \frac{1}{12}$.

The elementary rules for S and W offer an easy clue to the relationship between the two. Logarithm function is an obvious choice in view of the identity $\log a + \log b = \log(ab)$. Entropies being additive and probabilities multiplicative the relationship

$$S \propto \log W \qquad (10.10)$$

The entropy of a system is proportional to the logarithm of the probability of its state.

This relationship was first proposed by Boltzmann. It bridges particle statistics and macroscopic thermodynamics. The sheer fact that S and W increase and decrease in parallel reveals much about the molecular level content of the notion, entropy. A state is of high probability if the particles have a rich choice of possibilities to bring about a system of a given energy. Were each particle of exactly the same energy, the particles would have only one single way to add up the system energy, this being a very improbable state. If there is a rich choice of energies for the particles the possible combinations of energies are large: in that case the energy loss of a few particles can be compensated by the energy increase of a large number of low energy, or a smaller number of high energy particles. In short, the bigger the mess is in the particles' energy market, the more probable the state, the higher W is.

Disorder is not only a matter of energy. Let a mixture be made up of two substances, for example gaseous nitrogen and hydrogen. Complete order would correspond to the state where the positions of each molecule are fixed. In that case, there would exist only one single possibility for the spatial distribution of the molecules. Even a less strict order would limit probability. If H_2 molecules resided exclusively in one part of the vessel and N_2 only in the other part one could be sure which kind of gas molecules are grabbed once one knows in which part of the vessel she fumbles. Such separation in space, however, would hardly ever occur since the molecules moving at random would gather very-very rarely that way. The most probable state is, according of what is always experienced, that of the homogeneous, uniform mixture where one cannot have any idea whether a hydrogen or a nitrogen molecule is at a given position, offering a rich choice of the molecules' spatial arrangements.

To cut it short, entropy increases with disorder; it may even be regarded as the measure of disorder. That is the reason why it was stated earlier that the notions energy and entropy are very different on an atomic scale. Any single atom has a certain energy the energy of the macroscopic system being the sum of atomic energies. The entropy of an atom cannot be considered in a similar sense since entropy refers to the multitude of particles.

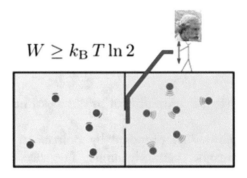

Fig. 28 Maxwell demon

This example might somewhat over-visualize entropy it being certainly something more than microscopic hide and seek. The high number of particles makes knowing each particle's position, path and energy impossible. We make do with the information that is sufficient for the atomic scale understanding of macroscopic events, those that are observed by temperature, mass and volume determinations. The fates of the particles are known only to the necessary depth, which is sufficient to understanding macroscopic observations. Beyond that limit, everything is left unknown. Entropy is the measure of that missing information. This is how entropy and disorder are connected: the positions of the particles are among the many unknowns. That is why information theory, an important branch of mathematics, applies the term and notion of entropy with a similar meaning. And that is the reason why Eugene Wigner called entropy an anthropomorphic notion, a human-shaped idea.

Maxwell devised a thought experiment to make clear the meaning of entropy and thermodynamic probability. As depicted in Fig. 28 a demon, able to distinguish and measure atoms one by one, guards the gas atoms. Initially, the gas temperature was the same in both compartments divided by a wall with a drop door. If a fast particle reaches the door from the left the demon opens it, thus enabling the particle to fly into the right hand side compartment. If a slow particle nears the door from the right the demon allows it to fly into the left hand compartment. Repeating this selection again and again the demon enriches the fast (high energy) particles in the right hand side, the slow (low energy) particles in the left hand side compartment. In other words, one of the compartments gets warmer the other one colder thanks to the demon's intervention, although initially they were of the same temperature. It is exactly of opposite what is expected by our daily experience and by the entropy law: from the state of thermal equilibrium the temperature difference between the compartments gets ever larger. Sure, because

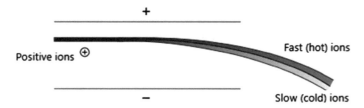

Fig. 29 Electric field separates the ions of different velocities

the demon observes the molecules one by one, information content increases hence entropy decreases.

Something similar can be done without demonic action, an electric condenser can do the job (Fig. 29). Electrically charged atoms (ions) of the same temperature are shot between two electrified plates. The electric field deflects the ions the slower ones being deflected more than the faster ones. Having left the field between the plates the temperature of the ion beam is not uniform any more: the slower ions are closer to the negative, the faster ones to the positive plate a temperature drop having been brought about in the direction perpendicular to the beam. This is no thought experiment, that kind of velocity selectors are often used in experimental ion physics, for example in mass spectrometry.

Two Nobelists of accelerator physics, *Rubbia* and *van der Meer* obtained the prize in 1984 for the construction of a "super Maxwell demon" if this is the right word. Ion velocities were not just measured in their equipment but the particles were accelerated and decelerated with appropriate electric fields so as to produce only one single energy in the gas which initially consisted of ions with a broad selection of energies. It means that the most improbable energy distribution of Fig. 27b was produced.

The marriage of mechanics and thermodynamics has turned out to be most fruitful. There was, however, an important problem, which caused no little headache. Talking about phenomenological thermodynamics it was stressed repeatedly that real processes are always irreversible. Thermal processes have a natural direction, i.e. a way they autonomously proceed, e.g. temperature differences disappear, different gases mix, and the opposite of those processes can be effected only at the cost of some surplus energy. Mechanical processes are, however, strictly reversible. How to explain the fact that whereas solitary atoms and molecules obey the reversible laws of mechanics, whenever they form a multitude they behave in an irreversible way?

Boltzmann was obviously aware of this question, his opponents saw the weakest point of his theory, being prone to refute the entire idea. The answer lies with the role of probabilities. Statistical mechanics is not solely based on mechanics it also contains the theorem of the spontaneous increase of probabilities. The latter, due to the relationship between S and W, is equivalent with the Second Law on entropy increase. Boltzmann solved the contradiction between mechanics and thermodynamics in the following way.

> The transition from the ordered to the disordered state is only most probable. The opposite has also a defined, even unimaginably small probability which approaches zero only as the number of molecules becomes infinitely large. [...] This, however, should not be imagined as if two gases, housing in a vessel of $\frac{1}{10}$ litre with completely smooth and inert walls, would be there initially unmixed, get mixed, after a few days become unmixed, then mix again, etc. [...] Only after a horribly long period of time, some $10^{10^{10}}$ years after the first mixing would appear de-mixing to a perceptible extent. [...] In practice that means never.

Indeed, an 0.1 L portion of a gas consists of a huge number of particles and probability, W is the larger the more particles make up the system. If, however, one chooses some rather small system, thermodynamic probability becomes small; hence, one may observe events that do not obey the law of probability increase, in other words the entropy law. As written above, it is most probable that hydrogen and nitrogen molecules fill a vessel in homogeneous and uniform distribution. Is this true even if the vessel is much smaller than 0.1 L? It is very difficult to produce an extremely small vessel and to take samples from different points. Instead of that it might be viable to measure extremely small parts of a vessel of any size.

A ray of light when traversing substances of different compositions or densities changes its direction; the effect is called refraction of light. Illuminating two gases before getting mixed light gets refracted at their boundary. As a reminder see Fig. 30.

Fig. 30 Refraction at the boundary of two substances

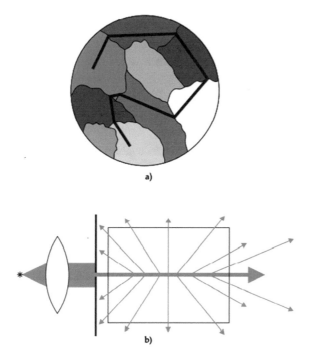

Fig. 31   a Light path in an inhomogeneous medium. **b** Light scattering

The wavelength of the visible light is much shorter than the size of any common lab vessel it being several hundred nanometres, that is several tenths of one millionth of one metre. If the gas mixture is inhomogeneous in such small regions (mixing is not perfect in volumes as small as several hundreds of cubic nanometres) then light is refracted at the boundaries of these small volume elements. Whenever light reaches a volume element which differs from its neighbour as far as composition goes, they ray changes direction (Fig. 31a). In a non-homogeneous medium light does not travel along a straight line but gets refracted at each site of local inhomogeneity. Hence, observing the macroscopic system as a whole, one sees light to exit the vessel also at any angle measured from the incident beam (Fig. 31b). That effect is called light scattering.

Sensitive methods show that light scattering takes place even in the purest, most carefully prepared mixtures. That is so because in volumes as small as the length of the light wave concentrations are not the same always and everywhere but change from point to point and moment to moment; concentration rises and falls, it fluctuates. Moreover, light scattering can also be observed in pure, one-component systems. Fluctuation of density, i.e. of the number of molecules per volume, can also be observed if the volume is small.

In volumes of the size of optical wavelengths the number of particles often differs from the average, in other words density changes from site to site. This explains why light scattering can also be observed in single-component systems. The investigation of small systems reveals the statistical, probabilistic nature of homogeneous mixing or uniform density. Homogeneity and uniformity are probable; the larger a system, the more so, but neither of them are ever certain.

Statistical calculations enable one to evaluate the expected fluctuations of concentration or density. The results of light scattering measurements can be compared with these theoretical results, calculations and experiments having been found in fair agreement. Those results render the most convincing proof of the basic idea of statistical mechanics: it is correct to regard substances as a multitude of molecules the motion of which can be described in terms of mechanical laws. Most naturally, the idea of probability necessarily involves deviations from the most probable behaviour. Macroscopic laws, first of all the increase of entropy, can be understood through the theorem of the increase of probability. The final proof of the atomic structure of substances was yielded by the observations of fluctuations, showing that all the basic tenets of the theory are of a statistical nature. Exception is the proof of the rule? Sure! The rules of statistics are proved by a well-defined number and extent of exceptions.

Often it is easier to create a new theory than to convince old-fashioned doubters. It is a peculiarity, or rather a tragedy, of science history that Boltzmann did not live to see the theory of fluctuations. As a result, many physicists regarded the statistical theory as one of the possible explanations at best, being unable to understand the appearance of the new law of nature. Because of the poor acceptance of his theory or perhaps due to some serious physical illness Boltzmann committed suicide at the age of 62. Four years later Einstein published his first paper on fluctuations.

The most prestigious of the opponents at the beginning of the twentieth century, *Wilhelm Ostwald*, who was a disbeliever even of the existence of atoms, let alone statistics, changed his mind only after Boltzmann's death. He became convinced by the good agreement of the values of the Avogadro constant obtained through a number of methods most of them based on the statistical theory. All expressions that contain the Boltzmann constant, k_B render the Avogadro constant in view of the known relationship $k_B = R/N_A$ = (gas constant)/(Avogadro constant). For example, the barometric formula or the above discussed fluctuations yield k_B, the ideal gas law gives R; the value for N_A is convincingly near to the figure obtained by Loschmidt who studied the viscosity and liquid state of gases.

11   On Molar Heat—Once More

It was quite a detour we made into basic quantum mechanics and molecular statistics. Our point of departure were the "anomalies" of the molar heats of perfect gases, anomaly meaning simply an observation that is unexpected at the current level of our knowledge and understanding.

The molar heat of gaseous hydrogen is $\frac{5}{2}R$ at constant volume and is independent of temperature, in agreement with the simple classical theory, whereas the molar heat of gaseous chlorine is different from $\frac{5}{2}R$ and depends on temperature. This is something that one would not expect since there is no difference between the structures of the H_2 and Cl_2 molecules within the framework of the classical theory. Nevertheless, equipartition law, basic as it might be, seems to be violated in the case of the chlorine molecule. Is it possible that H_2 molecules behave as a rigid dumbbell (Fig. 21), while Cl_2 molecules as an oscillator (Fig. 23), both obeying classical mechanics? No, this would give no answer to the problem of temperature dependence.

It is unpleasant to accept that simple substances of a similar appearance obey different rules; this makes it difficult to find underlying basic laws. Let us try to create a unified picture for the two molecules. The law of equipartition is strictly valid only if the energies of the molecules change in a continuous manner. Whenever molecular energies are quantized, equipartition is not effective any more. Strictly speaking, classical mechanics never applies to molecules; nevertheless, it is a reasonable approximation for a number of cases. Molecular energies are always quantized, there are only distinct energy levels with no continuous transition from one to the other, still discontinuities can be neglected in certain cases without making any serious mistake.

The molecular energy of translation (of free flight) was seen to be $\frac{3}{2}k_BT$— here classical and quantum mechanics are in accord. Rotations and oscillations are different in classical and quantum mechanics; still, provided the spacing of the quantized energy is much smaller than the translation energy, classical physics describes the fate of energy reasonably well. In that case, "the molecule cannot tell its own future", energy exchange is not influenced by the final state; hence, equipartition is maintained. In short, if inequality

$$(\varepsilon_2 - \varepsilon_1) \ll k_B T \tag{11.1}$$

prevails, classical physics is a good approximation.

In the opposite case, when the distances between the quantized levels are much higher than the average energy of translation,

$$(\varepsilon_2 - \varepsilon_1) \gg k_B T \qquad (11.2)$$

there is no energy exchange between colliding molecules because translational energy is insufficient to attain even the nearest quantized level. These are transparent situations. Things get more involved when the differences between the quantized energy levels are comparable with the average energy of translation,

$$(\varepsilon_2 - \varepsilon_1) \approx k_B T \qquad (11.3)$$

Energy exchange can proceed in such cases, but not according to classical mechanics. Hence, equipartition law is not valid. That is the reason why the molar heat of a gas depends on temperature. By varying the temperature the translational energy of the order of $k_B T$ changes and the conditions for the three above written cases are fulfilled one after the other.

A rough scheme of the vibration and rotation energy levels of the molecules Cl_2 and H_2 is given in Fig. 32. The distances between neighbouring vibration levels are always larger than those of the rotation levels. Rotation levels are usually much nearer each other than $k_B T$ around room temperature, hence rotation energies can be described by classical mechanics, equipartition law is valid. Vibration levels of H_2 are much further apart than $k_B T$; thus, vibrations cannot be excited by thermal motion and the molecule behaves as a rigid dumbbell resulting in the molar heat $\frac{5}{2} R$, independent of temperature (in technical parlance the vibrations are frozen). The vibration levels of Cl_2 are much nearer to each other, they can be thermally excited, the easier the higher the temperature is. At a low temperature Cl_2 is also similar to a rigid dumbbell; hence, the molar heat of the gas is $\frac{5}{2} R$. As temperature increases the molecule oscillates more and more intensely; finally, as $k_B T$ gets much larger than the energy distances of the vibration states, they obey classical mechanics, with molar heat approaching $\frac{7}{2} R$.

That conclusion is predicted by theory and proved by measurements. Molar heats of gases are but a minor problem in physical chemistry. We can gather, however, a general insight from the above considerations. Atomic, molecular processes can be described only in terms of quantum mechanics correctly. Still, there are certain cases and conditions for which classical mechanics render reasonable approximations. It is difficult to give general rules here. However, common sense is a good tool in physical chemistry (as well).

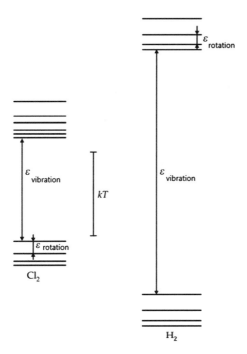

Fig. 32 Scheme of the rotation and vibration energy levels of the Cl_2 and H_2 molecule

12   The Hydrogen Atom—Is Something Revolving and Spinning?

On occasion, molecules were imagined as elastic spheres; at other times, as rigid dumbbells or vibrating springs, as discussed on the previous pages. The theoretical results based on these models, various as they were, turned out to be reassuring insofar as they agreed with the experimental facts. It is, however, worrisome that molecular models were devised according to the given problem to be solved and were not based on firm physical principles. This despite the fact that it is clear that quantum mechanics is the tool to reveal the structures of atoms and molecules. If we are going to understand the structure of substances in a greater depth, as far as both exactitude and the behaviour of the smallest constituents of matter are concerned, we have to apply the laws of microphysics to the structure of molecules.

Luckily enough, there is no need to start from scratch. It is common knowledge (this book supposed it already in the previous chapters) that an atom consists of a positive nucleus and revolving negative electrons. Revolving? Some caution is in order here. Earth revolves around the Sun along an elliptic path its position and velocity being known at any moment. Talking about "new physics", we stressed that electrons have no paths in that sense of the word. It is only the probability of their position that is predicted by the theory: how often they can be found in the neighbourhood of a given site. That probability is given as $|\psi|^2$, the square of the absolute value of the eigenfunction. If we want to visualize the electron around the nucleus the word "path" is certainly misleading, "electron cloud" being a more appropriate expression. Electrons surround the nucleus appearing at different positions with different probabilities, this can be expressed by saying that the electron cloud is denser at a certain place and thinner somewhere else. Microphysics does not know anything about planetary paths or fast moving, electrically charged peas.

The notion of the eigenvalue equation and the physical content of their solutions were described earlier. At the outset of the treatment of a quantum mechanical problem one has to construct the energy operator, $\hat{H}$. The energy of a macroscopic body is the sum of its kinetic and potential energy; similarly, the energy operator is the sum of the particle's kinetic and potential energy operators. In order to construct it, one has to know the constituents of the atom.

At the beginning of the last century, Rutherford made a most important discovery. Having bombarded of a thin layer of some material with α particles he observed that some of the particles were deflected or even recoiled. In that time α particles were already known to be identical with swift moving, positive nuclei of He atoms. How come such a fast particle is recoiled from a very thin layer? Rutherford himself was greatly astonished: "*It was almost as incredible as if you fired a 15-inch shell at a piece of tissue paper and it came back and hit you.*"—he said. He devised an explanation, stating that the density of the layer is not uniform but at certain sites it is very high (large masses in small volumes); moreover, these large masses carry positive charges. Incoming positive α particle and positive charge in the layer repel each other, whereas the large mass greatly deflects the bombarding particle. Heavy, positive centres within the substance: this experiment and its interpretation was how the atomic nucleus was discovered. Rutherford's experiments also gave the amount of positive charges in the nucleus. Consequently, the atom being neutral, also the number of electrons around the nucleus.

A hydrogen atom consists of one electron and a singly charged positive nucleus. Thus, three energy contributions are to be considered for the operator $\hat{H}$: the energy of attraction between nucleus and electron, the kinetic energy of the electron and the kinetic energy of the nucleus. Figure 33 is just a reminder to the contributions forgetting for the sake of simplicity that the electron is no tiny ball.

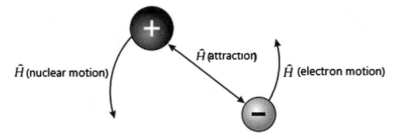

Fig. 33 Scheme of a H atom—the electron is no tiny ball in reality

Mass spectrometry and other methods show the mass of the nucleus to be some 2000 times that of the electron, hence the nucleus is practically at rest, its kinetic energy can safely be neglected. That makes the energy operator of the H atom to be approximately as

$$\hat{H}(\text{H atom}) \cong \hat{H}(\text{attraction}) + \hat{H}(\text{electron motion}) \tag{12.1}$$

That operator figures in the eigenvalue equation,

$$\hat{H}(\text{H atom}) = E \cdot \psi \tag{12.2}$$

The solutions of this equation render the possible energy levels of the H atom, together with a series of ψ functions, the absolute squares of which are the probabilities of electron distribution around the nucleus. The solution is "almost only a mathematical problem", "almost" because physical reality must be always considered and not "only" because the treatment is much involved. Nevertheless, it can be performed and the result is among the first great achievements of quantum theory.

The optical spectrum of the H atom reveals the atom's energy levels because, as it was discussed before, the relationship

$$(\text{difference between two atomic energy levels}) = 2\pi\hbar \cdot (\text{optical frequency}) \tag{12.3}$$

Let the initial energy of the atom be $\varepsilon_{\text{initial}}$ which is decreased to $\varepsilon_{\text{final}}$ due to light emission; the energies and emitted light frequency are interrelated as

$$\varepsilon_{\text{final}} = \varepsilon_{\text{initial}} - 2\pi h\nu = \varepsilon_{\text{initial}} - \hbar\omega \tag{12.4}$$

Recalling the fundamental relationship between energy and frequency, this is nothing else but energy conservation law applied to light emitting and

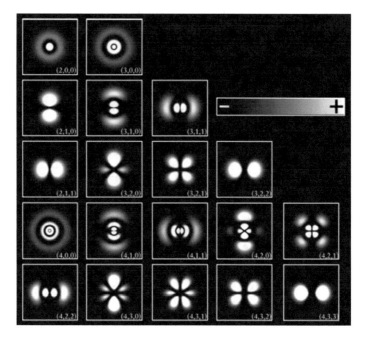

Fig. 34 Electron density distributions for different H atom energies. The first figure in parenthesis gives the energy level (based on Wikipedia)

absorbing atoms. By measuring spectral frequencies, atomic energy levels can be found.

The solution of the eigenvalue equation renders electron energies, expressed as

$$\varepsilon_n = -\frac{m_e e^4}{2\hbar^2 n^2} \tag{12.5}$$

where m_e and e denote electron mass and charge, respectively, and n is any positive integer. The formula predicts a series of distinct energy levels which are nearer to each other as the energy gets higher.

Results of experimental spectroscopy are in convincing agreement with these calculations. As a matter of fact, the basic laws of quantum mechanics were revealed by Heisenberg through studying the optical spectrum of the H atom.

A number of computed probability distributions of electron around the H nucleus are given in Fig. 34. Electron densities, i.e. probabilities of finding an electron in a certain region around the nucleus, are depicted for several energies. Some of these shapes are surprising since the radial Coulomb force of attraction would suggest spherical symmetry. There are, however,

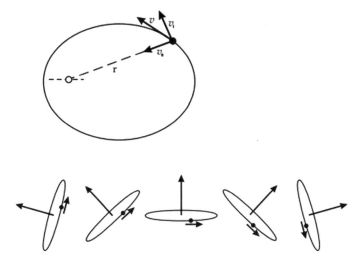

Fig. 35 To the question of planetary motion

several distributions with, for example, cigar-like or clover shapes. The central symmetry of force is still visible also in these cases, the nucleus being always in the symmetry centre of the distributions.

It is remarkable that more than one density distribution might correspond to a given energy—that is, one eigenvalue might be accompanied by a number of eigenfunctions. It is perhaps not too surprising that there are analogies in classical physics: the potential energy of a pound of potato on the third floor is much the same whichever way the tubers are scattered on the kitchen floor. Energy, however, is not the only quantity that characterizes the motion of a body, neither in classical, nor in quantum mechanics.

An early problem of classical mechanics, that of planetary motion, is recalled in Fig. 35. The kinetic energy of a planet moving along its elliptic path is $\frac{1}{2}mv^2$ with m being the mass and v the velocity of the planet. Velocity being a vector, it can be decomposed into two components, one pointing toward the Sun, $v_{\parallel}$ and the other one perpendicular to it, $v_{\perp}$. As the planet moves on its velocity varies and so does its potential energy, U_{pot} because the distance from the Sun changes. Neglecting all other forces but the Sun's attraction energy, conservation reads as

$$\frac{1}{2}mv_{\perp}^2 + \frac{1}{2}mv_{\parallel}^2 + U_{pot} = E = \text{constant} \qquad (12.6)$$

The first member is particularly important in the mechanics of revolutions. The vector, called angular momentum, L being a function of $v_\perp$ and the distance from the centre of revolution, r, is defined as

$$L = mrv_\perp \tag{12.7}$$

Angular momentum remains constant throughout revolution, provided attraction is the only acting force, similarly to the momentum which remains constant throughout translation, provided no force acts on the moving body. Energy can be expressed also by making use of L as

$$\frac{L^2}{2mr^2} + \frac{1}{2}mv_\parallel^2 + U_\text{pot} = E = \text{constant} \tag{12.8}$$

True, both L and E are constants but infinitely many L values might go with a given energy. Different L values are balanced by different contributions of the two further quantities, so as to maintain E constant. Revolution with a certain energy can exist with an infinite amount of different angular momenta in classical physics. Moreover, angular momentum is a vector with a given orientation. Two revolving systems, even with the same energies and the same magnitudes of angular momenta, might differ as far as the planes of revolution are concerned, as it is drawn in Fig. 35.

Having visualized some of the notions of classical physics by planetary motion, it is obvious that we also need to consider the rotation of a sphere around its own axis. Since each particle of the rotating sphere performs a circular motion, there exists an angular momentum also of rotation, L_rot this being also constant. The energy of rotation is again proportional to L_rot^2 and adds to the total energy of the system. The absolute magnitude of L_rot might still be the same whereas the orientations of the rotation axes are different (Fig. 36).

Angular momentum is one of the basic quantities of classical physics, which was both necessary and possible to be transplanted into quantum mechanics.

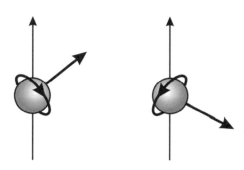

Fig. 36 Various angular momenta belongs to one single energy of rotation

Despite that, do path, revolution, rotation have no meaning there? Sure they do, because energy, momentum, force, distance do retain their meanings in terms of which angular momentum is defined. A quantity with the dimension of the angular momentum does exist in quantum mechanics. (The word dimension was explained in connection with the universal gas constant, R in the chapter on perfect gases.) The relationship between mass, force, velocity, distance and angular momentum is the same both in classical and quantum mechanics, angular momentum being also conserved in quantum mechanics. However, planetary paths, rotating balls must not be associated with all these quantities, as they are unknown in quantum mechanics.

Something else is also well known. The figures in Fig. 34 represent electron densities. According to the first numbers in parentheses, the figures are related to certain energy states. Different electron densities correspond to angular momenta, different in magnitude and orientation. One thing is missing from the figures and that is the angular momentum related to rotation in classical physics, which is called spin in quantum mechanics.

Detailed calculations yield the number of angular momenta that go with a given energy. Here the difference between classical and quantum physics becomes most clear. In classical physics, L can take whatever value, similarly to energy. In quantum mechanics, both can vary only discontinuously, in jumps. Angular momentum changes in steps the extent of which being at least $\hbar$. To cut it short, there is one angular momentum for the lowest energy state of the H atom, four for the second, nine for the third state.

Spin states, being akin to the rotation of a ball, can have only two values: $\frac{1}{2}\hbar$ or $-\frac{1}{2}\hbar$. As a result of that 2 different angular momenta belong to the lowest energy state of the hydrogen atom, 8 to the next and 18 to the third, different angular momenta meaning also different electron densities. These numbers will soon turn out to be important.

13   Atoms—Electrons Are Unaccommodating

It was an obvious demand to apply the method, developed for the hydrogen atom, to the spectra of larger atoms. Nuclear charges and the number of electrons having been known through mass spectrometry and similar methods, the task was to find the energy operator, $\widehat{H}$ and to solve the eigenvalue equation.

The helium atom problem is sketched in Fig. 37. The positive nucleus having two elementary charges with two electrons around it represents a problem more complicated than that of the hydrogen atom because electron–electron repulsion is also to be accounted for. Neglecting nuclear motion, the

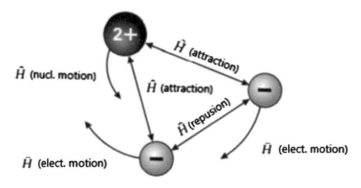

Fig. 37 Sketch of the helium atom (the electron is no tiny ball in reality)

energy operator is composed of the terms below:

$$\hat{H}(\text{He atom}) \cong \hat{H}(\text{attraction between nucleus and one of the electrons})$$
$$+ \hat{H}(\text{attraction between nucleus and the other electron})$$
$$+ \hat{H}(\text{electron} - \text{electron repulsion})$$
$$+ \hat{H}(\text{motion of one of the electrons})$$
$$+ \hat{H}(\text{motion of the other electron}) \qquad (13.1)$$

Mathematical discussion is becoming more involved than before since the energy operator is more complex. A multitude of eigenvalues and eigenfunctions are expected. True, but less of them is to be found than one would think without considering the physical content of the equation. Independent operators were constructed for each of the electrons hence the probability densities of each electron around the nucleus could be evaluated. The result would show that one of the electrons is mostly here, the other one is there. But how to tell the two electrons apart? Two balls of the same size and colour rolling on a billiard board can be distinguished by their distinct paths: at a given moment one of them is here, the other one there. But quantum mechanics cannot say anything about paths, hence electrons of the same mass and charge are indistinguishable. We can tell the probability of finding electrons in a certain volume element but it is meaningless to ask which of the electrons is there.

Thus, eigenfunctions must not give the state of one or the other electron but that of the electrons as a whole. As far as experiments are concerned, only $|\psi|^2$ is meaningful, this rendering the occurrence probability of the electrons. Let electron 1 be in state m and electron 2 in state n. The simplest of the propositions would be to try the product.

$\psi_m(1)\psi_n(2) = \psi_{\text{perhaps}}$ for the eigenfunction of the two-electron system, since eigenfunctions are related to probabilities and the product of independent probabilities does have a meaning as it was discussed in connection with entropy.

Not here! This "perhaps" function would tell more than reasonable since it tells that electron 1 is in state m and electron 2 in state n. Moreover, it is not the eigenfunction but its absolute square that expresses probability. Some more complicated eigenfunction must be found. Eventually this:

$$\psi_s = \psi_m(1)\psi_n(2) + \psi_m(2)\psi_n(1)? \qquad (13.2)$$

Or this:

$$\psi_a = \psi_m(1)\psi_n(2) - \psi_m(2)\psi_n(1)? \qquad (13.3)$$

Function ψ_s is called symmetric, ψ_a antisymmetric eigenfunction. Both of them express the fact that the electrons cannot be discriminated, probability densities $|\psi_a|^2$ or $|\psi_s|^2$ remaining the same if numbers 1 and 2 in parentheses are exchanged. Right, but do both of them describe He electrons or more generally, the electrons of any atoms appropriately?

Only experience can give an answer here. As a critical problem, let us investigate the functions with both electrons in the same state, $\psi_m = \psi_n$. The symmetric function behaves in a simple manner,

$$\psi_s(\text{two particles in the same state}) = 2\psi_n(1)\psi_n(2) \qquad (13.4)$$

The antisymmetric function yields a much different result,

$$\psi_a(\text{two particles in the same state}) = \psi_n(1)\psi_n(2) - \psi_n(2)\psi_n(1) = 0 \qquad (13.5)$$

The antisymmetric eigenfunction is zero, so is obviously its square; the probability of two particles being simultaneously in the same state is zero. Two particles in the same state, that is simply impossible—provided the antisymmetric function is the correct one.

As said, an experiment must decide. If experiments show two electrons to exist in the same state, than their behaviour is described by an eigenfunction of ψ_s type; if such a situation does not exist, ψ_a is the valid function.

Performing Rutherford-scattering experiments over the chemical elements it turns out that increasing nuclear mass is accompanied by an increasing number of positive nuclear charges and obviously with an increasing number of electrons. There is a long line starting from the one-electron hydrogen and the two-electron helium to the ninety-two electron uranium. The unfathomable wealth of the physics and chemistry of inanimate and living nature comes from the properties of the ninety-two elements. It will be discussed in more detail but it is common knowledge that the constructions of the atomic electron shells determine the chemical properties. Similarly, also the optical spectra stem from the electrons. One may toy with the question what nature would be like, including ourselves, if the electrons obeyed symmetric or antisymmetric wave functions.

Electrons tend to attain states of the lowest energies, like the objects on the Earth which move towards the centre of the globe trying to find the place of the lowest potential energy, provided they are not prevented from doing so. Could any number of electrons be in the same state, atoms would differ only in the number of electrons whereas their energies and densities would be approximately the same. Only electron–electron repulsions would increase with increasing number of electrons. This would result, however, in a minor effect, particularly with the heavier atoms: the difference between the forces exerted by 53 or 54 electrons cannot be very large. If all the electrons were in the same state, iodine crystals would be very similar to xenon gas.

Nature is experienced to be much different. Atomic spectra and properties of the elements are much richer than promised by the imagined world of symmetric eigenfunctions. Moreover, the landscape changes periodically. No, properties do not vary in parallel with the number of electrons and many electron atoms do not resemble each other. What is observed is the fact that atomic masses and nuclear charges play but little role in the properties of the elements. The basic law of periodicity was revealed by Mendeleev almost sixty years prior to modern physics.

If one counts the electrons of elements which are similar in their physical and chemical properties, one finds rules that may make a numerological impression. The number of electrons of similar, lighter elements differ by eight. For example, in the rare gas group He, Ne, Ar there are 2, 10 and 18 electrons, respectively, whereas 3, 11, 19 electrons are in the alkali metal group Li, Na, K. Going to heavier atoms the differences are 18 with even more heavier ones becoming 32: the three heavier rare gases Ar, Kr and Xe have 18, 36 and 54 electrons, three halogens Cl, Br and I have 17, 35, 54 the differences between the neighbours being invariably 18. The electron numbers in the very similar Ba and Ra are 56 and 88, respectively, whereas in Cd and Hg, two elements being much akin, they are 48 and 80 the differences being for both pairs 32. Casting a glance at the Periodic Table, further examples are easy to find.

A periodic and most multifaceted world seems to be impossible to explain in terms of symmetric eigenfunctions.

Antisymmetric eigenfunctions demand a more varied selection of electron states. When a new electron enters the cloud around the nucleus it must find a state the ψ function of which differs from those of all the electrons already present. It is clear that antisymmetric eigenfunctions promise a world much richer in a variety of phenomena than symmetric eigenfunctions could do. Still, periodicity of the properties is not a necessary consequence of this consideration.

Periodicity, the structure of the Periodic Table is explained by the hydrogen atom. Multiply charged positive nuclei with a correspondingly high number of electrons resemble the hydrogen atom only faintly. Still, with only one electron missing from an atom, the remaining atomic core has the same amount of charge as the hydrogen nucleus. This last electron does not differ too much from the hydrogen's one and only electron. At least as far as the number of states at a given energy is concerned. This counting can be done in the same way as in the hydrogen atom case the solution of the hydrogen eigenvalue equation being made use of.

If the electron eigenfunctions are antisymmetric, two electrons are impossible to be in the same state, hence as many different states must be taken as there are electrons present in the atom. Analogy with the hydrogen atom tells the numbers of states having been given at the end of the previous chapter. Let us look at the beginning of the Periodic Table (Fig. 38)!

The lowest energy state can take two electrons at most hence two electrons of very similar states must be present around the helium nucleus. The third electron around the triply charged lithium nucleus must find a state of much higher energy this, however, can house seven further electrons provided

I	II	III	IV	V	VI	VII	VIII
$_1$H							$_2$He
$^1_2$L	$^2_2$B	$^3_2$B	$^4_2$C	$^5_2$N	$^6_2$O	$^7_2$F	$^8_2$Ne
$^1_8$Na $_2$	$^2_8$Mg $_2$	$^3_8$Al $_2$	$^4_8$Si $_2$	$^5_8$P $_2$	$^6_8$S $_2$	$^7_8$Cl $_2$	$^8_8$Ar $_2$

Fig. 38 Beginning of the periodic table by Mendeleev. The numbers of electrons of each shell is given

their angular momenta, including their spins, are different. That is how the shells of the atoms, in the order of their increasing masses, are populated reaching neon, the second of the rare gases. The extra electron of sodium, the 11th element, must be of a higher energy than any of those of the previous eight elements. From here on the third row of the Periodic Table is being constructed, ending with the rare gas argon. Reaching the end of the row the states in the shell are all filled. Next to the symbols of the elements the number of electrons in different states but of (approximately) the same energy is given in Fig. 38.

Looking at the number of electrons it is reassuring to meet the magic numbers discovered by Mendeleev as he constructed the Periodic Table based upon chemical experience and his brilliant intuition. May we repeat, these numbers are the amount of electrons whose energies are approximately equal. (These energies are strictly the same in the hydrogen atom. The electron–electron interactions in other atoms alter the energies slightly without influencing the number of possible states.)

As a matter of fact, we are extremely lucky, since all atoms in nature are built according to the pattern of the hydrogen atom. The simplest of the cases renders the solution of the general problem.

The periodicity of properties is now almost self-evident. Those elements are similar to each other whose electron structures are similar, that is, have the same number of electrons beyond the closed shells. They make up a column of the Periodic Table.

One of the most basic and general observations in chemistry was explained in a simple and pictorial manner by assuming that electron eigenfunctions are antisymmetric. No similar result could be achieved by making use of symmetric eigenfunctions. Experimental facts compelled us to accept the antisymmetric functions as the correct choice.

Frankly spoken, physicists have further proofs although the structure of the Periodic Table is an important argument. It turned out that if a particle's spin is $\frac{1}{2}\hbar$ or $-\frac{1}{2}\hbar$ or their integer multiple, the eigenfunction is antisymmetric; if the spin is $\hbar$ or its integer multiple, the eigenfunction is symmetric. The latter case is valid for an α particle among numerous other ones. Spin is a measurable quantity; it is needless to say that the electron spin was invariably found to be $\frac{1}{2}\hbar$.

The law regarding the spin states and the compatibility of particles, called the Pauli principle, holds for every microphysical particle. It states that if the particle spin is an integer multiple of $\hbar$ its eigenfunction is symmetric thus any number of particles can have the same state; whereas if the spin is an integer multiple of $\pm\frac{1}{2}\hbar$ the eigenfunction is antisymmetric, thus barring two particles to be in the same state.

14   The Hydrogen Molecule—To Have Some Elbow Room

It has been well known since the age of Avogadro and Cannizzaro that gaseous hydrogen, similarly to all gases excepting the rare ones, consists of diatomic molecules. Obviously so, because experiments show that the aggregate energy of two hydrogen atoms is smaller if united than if moving independently. Why does the energy decrease if two atoms unite?

The simple answer is: attraction. Atoms are made of attracting positive and negative charge carriers. Well ... almost. Let us have a closer look!

Some compounds, like table salt, NaCl exist indeed due to electrostatic attraction. The constituents, sodium and chlorine are present in the crystal as electrically charged particles. That is indicated by the increase of electric conductance of water if table salt is dissolved an effect being due to the presence of Na^+ and Cl^- ions in the solution. Let us consult Fig. 38! If a Na atom loses one electron, forming an Na^+ ion, its electron shell becomes similar to that of the rare gas, Ne. If a Cl atom takes an electron, forming Cl^- ion, its electron shell becomes similar to that of another rare gas, Ar. These electron configurations being very stable (i.e. having low energies) are prone to be formed through the exchange of one or two electrons. The two ions of opposite charges are held together by electrostatic forces indeed.

It would be difficult to understand the formation of the hydrogen molecule in similar lines. Why should a H atom give an electron to another one? Built of two identical atoms, the molecule is expected to be completely symmetrical, with the constituent atoms playing equivalent roles. Let us try to follow up the formation of a valence bond between two identical atoms.

The hydrogen molecule, H_2 consists of two H^+ nuclei and two electrons. Nuclei and electrons attract, whereas the two electrons and the two nuclei repel each other. One might proceed similarly to the method described in the He atom case by composing the $\hat{H}$ operator of the molecule from the potential energies of the attractions and repulsions and the kinetic energies of the four particles. But our mathematician friends would face an insurmountable task by that. We have an easy way in these pages since it is not our aim to perform the calculations only to indicate the problems. The actual equations, however, are very involved, the H_2 problem cannot be solved exactly at all. One has to take recourse to approximations based also on one's graphic insight and physical intuition.

The basic principle referred to is that of the minimum energy. The constituents of the molecule are arranged so as to make their aggregate energy to be the lowest. Energy has no absolute zero level, the reference energy is well known to be a matter of convention. Now, the zero energy is chosen as the sum of the energies of the particles that are at infinite distances from one another so as to have no interaction. By making use of this zero level, the energy of a H atom, $-E_H$ is negative because energy is released if an electron starting out from infinity reaches the lowest energy shell. The aggregate energy of two H atoms which are infinitely far from each other, hence being unable to pay any heed to one another, is $-2E_H$. If two atoms were so near each other that their nuclei would be almost fused the electrons would experience an electrostatic field similar to that in the He atom: in both cases electrons are attracted by twofold positive elementary charges. At the same time, the two nuclei being at a mutual distance r have an electrostatic energy $\frac{e^2}{r}$ where e denotes the elementary charge. The energy of that "almost He atom" is

$$E^I = -E_{He} + \frac{e^2}{r} \qquad (14.1)$$

Here $-E_{He}$ denotes the total energy of the really existing He atom, it being equal with the energy of the two electrons in the atom. The aggregate energy of two independent H atoms is

$$E^{II} = -2E_H \qquad (14.2)$$

Detailed calculations on the He atom can be executed exactly; the results show that $-E_{He}$ is much smaller (larger absolute value negative number) than $-2E_H$ in accordance with the high chemical stability of the rare gases. That is the main point: the aggregate energy of two electrons in the neighbourhood of the doubly charged nucleus is much lower than the sum of the energy of two H atoms. Therefore, as two H atoms come closer and closer to each other the electrons' energy decreases, the smaller distance is advantageous; it is only the repulsion between the nuclei that limits the nearing.

Let the conditions be plotted in a coordination system with the internuclear distance on the horizontal, the total energy on the vertical axis (Fig. 39). The inter-nuclear repulsion, $\frac{e^2}{r}$ appears as a hyperbola in the upper, positive part of the graph. The negative, attractive contribution of the total

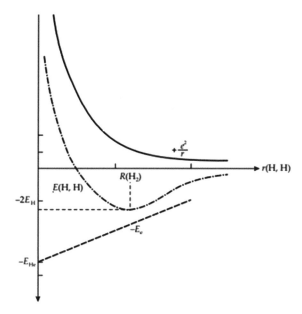

Fig. 39 Energy of the H_2 molecule as a function between two H atoms as it is approximated by making use of the atomic energies of H and He. The H–H distance corresponding to the energy minimum is the real bond length

energy, E_e changes between $-E_{He}$ and $-2E_H$, here just for the sake of simplicity it is assumed that it is proportional with the distance, as it is plotted in the lower, negative part. The total energy of the H, H system is the sum of the two contributions,

$$E(H, H) = -E_e + E_{nucl} \qquad (14.3)$$

The result appears as a chain-dotted line in the figure.

Even this rough calculation shows the existence of an inter-nuclear distance where $E(H, H)$ has a minimum. The aggregate energy of the two H atoms is here the smallest, this most stable state representing the H_2 molecule with the corresponding r value as the bond length, $R(H_2)$.

The chemical bond in the H_2 molecule was understood in terms of the conditions which prevail in the He atom. This is meant in a strict sense including the limits stipulated by the Pauli principle which bars two electrons to have the same spin: if one is $+\frac{1}{2}\hbar$ the other one must be $-\frac{1}{2}\hbar$. In short this is expressed by saying that the two spins are of opposite orientation or anti-parallel and by denoting as ↑↓. This is a necessary condition of a chemical bond; were the spins of the same orientation, parallel, ↑↑, then all properties of the two electrons would be the same and this is not tolerated by the

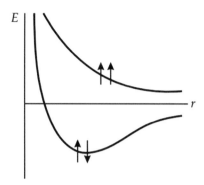

Fig. 40 The energy of the H,H system with parallel (↑↑) and anti-parallel (↑↓) spins

Pauli principle. The upper curve in Fig. 40 shows the aggregate energy of the two H atoms in case of parallel spin orientation. No minimum appears as a function of inter-nuclear distance energy getting never lower than $-2E_H$; the incompatibility of the same spin electrons are added to the nuclear-nuclear repulsion.

The H_2 molecule is the archetype of the covalent bond. Whenever valence and valence bond is mentioned and a straight line is drawn to symbolize the coupling of two atoms a covalent bond is meant. One encounters most frequently with the H–H, C–H, C–C, O–H with innumerable further ones.

There is an important, common property of all these covalent bonds which might reveal their essence perhaps better than any involved mathematics of the eigenvalue equation. Let us take recourse to the analogy of a swinging chord. The longer a chord is the lower its basic frequency; on a musical instrument, a longer chord sounds deeper. As discussed before, energy and frequency are proportional to each other if the chord is a microphysical object; thus its energy decreases with increasing length. Mathematical treatment shows that an electron localized in a molecule is similar to a chord as far as energy and allotted space are concerned. Larger space brings lower energy. Space being always larger in a molecule than in an atom both bonding electrons enjoy a two-atoms large volume hence their energies are lower than in the individual atoms.

Electron distribution is always symmetric between identical atoms. There is no reason for the electron to distinguish between two H or two C atoms electron densities being the same around the two atoms. If the atoms are different this strict symmetry breaks down. Nuclear charges and electron densities being different, the bonding electron appears with higher probability in the vicinity of one of the atoms than of the other. For example, in the O–H bond, electron density around O is much higher than around H.

As an extreme case, one of the atoms presents a much more advantageous environment than the other localizing the bonding electron practically there: that is the case of the ionic bond discussed by the example of NaCl.

The valences of the atoms, different as they are, can be understood by considering the construction of the Periodic Table. The highest number of electrons an atomic shell can accommodate equals the number of the electron states in a hydrogen shell: that consists of states having the same energy but different angular momenta. The maximum valence of an atom is the number of electrons that can saturate its outermost electron shell completely making it similar to a rare gas. The difference between the highest and the actual number of electrons gives the number of available sites on this shell. One electron turns the shell of H similar to that of He; four electrons make the shell of C, three electrons that of N to Ne shell, one electron transforms the Cl shell into one like Ar.

Being somewhat familiar in inorganic chemistry one certainly knows that there are elements like e.g. sulphur, phosphorous, iron or manganese which have more than one definite valence. Moreover, there exist so-called complex compounds, like the yellow prussiate of potash, $K_4Fe(CN)_6$, the structures of which are difficult to understand in terms of the traditional notion of valence. These problems are not to be treated here, let the reader just be reminded that the higher energy H shells can accommodate more than 8 electrons, there being place for 18 and further for 32 ones. Moreover, the idea that the electron structure of each atom can be deduced from the possible H atom states is an approximation only, inappropriate for the understanding of all the finer details. Even less can one expect that the molecular properties should be predicted completely and correctly in that way.

15   Different Kinds of the Hydrogen Molecule?

Rather not! Two H atoms being alike can be coupled only in a single way. Still, the spectrum of the hydrogen molecule raises some doubt. Whereas the spectrum of the H atom is independent of temperature, that of the H_2 is temperature-dependent. No wonder, since the rotations and oscillations of the molecule are thermal processes: if the gas is warmer these movements are more intense influencing the emission or absorption of electromagnetic radiation, that is, of light. Quantum mechanical calculations describe those processes correctly. Refined measurements, however, suggest that possible movements of the molecule have been forgotten.

That forgetfulness stems from our picture of the hydrogen atom. It is simply not true that there exists only one single kind of H atom, referring here to the lightest atom of atomic mass one. (Other complications will be discussed later.) Indeed, in the absence of any external force they cannot be told apart. Magnetic field, however, can reveal a difference. Let a beam of H^+ ions be passed through a strong, inhomogeneous magnetic field, inhomogeneous meaning the field strength to vary from point to point, the beam is separated into two weaker beams of equal intensities (Fig. 41).

It seems, magnetic field makes a property manifest, not to be observed otherwise.

Resting charges do not feel any effect of a magnet, a magnetic field acting only on moving electric charges with a force that is proportional to the velocity of the charge. Angular momenta of rotating or revolving particles are proportional with their velocities as it was already described. Thus, if the magnetic field has an effect on a particle, this shows that the particle does have an angular momentum. The sheer fact that a beam of H^+ ions, being just bare atomic nuclei, are split into two beams is a demonstration of the fact that the nuclei have two different angular momenta. This difference being

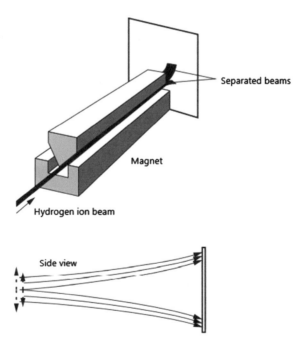

Fig. 41 A strong, inhomogeneous magnetic field splits the H^+ ion beam into two

non-perceptible in the absence of any magnetic field, the energies of the two kinds of nuclei must be equal.

Little attention was paid to the properties of the nucleus in the previous treatment of the H atom. It was simply stated that the nucleus has one positive charge and a mass so high that its motion does not deserve any particular attention. The above beam experiment revealed the angular momentum, the spin of the nucleus. Spin goes with charge motion, thus explaining the interaction with the magnetic field. The hydrogen atom nucleus consists in one single proton, an elementary particle. The above experiment shows that it is a half-spin particle similar to the electron, its angular momentum being $+\frac{1}{2}\hbar$ or $-\frac{1}{2}\hbar$. The absolute values of the two angular momenta being equal and differing only in their orientations, they cannot be perceived unless there is some effect that defines an orientation indicating which way is up and which down.

An inhomogeneous magnetic field is an effect like that. The varying density of the lines of force define the orientation. Another similar effect is the encounter of two H atoms where the relative orientations of the two nuclear spins reveal their existence. Hence two kinds of H_2 molecules can form from H atoms. The orientations of the nuclear spins are either the same, ↑↑, this being called ortho-hydrogen, or they are opposite, ↑↓, called para-hydrogen. The energies of the two kinds differ but only to a small extent. Still, it is measurable, if the relative amount of the two variants is studied. If the temperature is low, not much above 0 K, the proportion of the para-hydrogen is 100%. With increasing temperature, the contribution of the ortho variant increases becoming near room temperature around three times more than its counterpart. That is the maximum value, the proportion (3:1) being set by the angular momenta of the two variants. However, the gas synthetized at a low temperature reaches that maximum only slowly. The presence of metallic palladium or platinum surfaces enhances the process by decomposing the molecules into atoms which re-form ortho- or para-hydrogen molecules now in a proportion that corresponds to the actual temperature.

The sluggishness of the conversion is connected to a basic law of mechanics, both classical and quantum. Angular momentum in a closed system is constant; thus, a para-molecule must obtain some angular momentum to become an ortho-molecule. Sure, the metal surface accelerates the conversion due to the reactions $p\text{-}H_2 \rightarrow 2H \rightarrow o\text{-}H_2$, but also its atoms lend the missing angular momenta to the para-molecules.

Today the difference between the two variants of molecular hydrogen is of practical importance. Hydrogen fuelled space ships use mostly p-hydrogen because its evaporation loss is much lower than that of the o-version.

16   Different Kinds of the Hydrogen Atom?

Nuclear spin as realized in the chemistry of the hydrogen molecule has taken physical chemistry to the borders of nuclear physics. This was not the first of the effects by which chemists were warned not to limit their attention to the outermost shell of electrons. Right upon the invention of mass spectrometry researchers observed that pure elements being homogeneous as far as chemical properties are concerned consist of atoms of different masses. Surprising or not, it became obvious that atoms of different masses can have exactly identical chemical properties. For example, hydrogen, present in nature, contains an atom in a proportion of 0.016% which is twice as heavy as the overwhelming component H. This substance whose chemical properties agree with those of H was called deuterium, "second" (a not too witty naming) and is symbolized by D or ^{2}H. Increasing the sensitivity of methods, the existence of a third component, three times as heavy as H, present in a proportion 10^{18}:1, was also demonstrated. It is called tritium, a β emitting radioactive nucleus, its symbol being T or ^{3}H.

Similarly to hydrogen, most of the elements consist of atoms of identical chemical properties but of different masses. The effect is called isotopy, the atoms with differing masses of one element are the isotopes of each other. The words of Greek origin meaning "same place" refer to the Periodic Table where elements are ordered according to their chemical properties, hence isotopes of an element must share one and the same place.

The model-like explanation of isotopy is most visual. Nuclei consist of positive protons and electrically neutral neutrons. Charge of a nucleus is defined by the number of protons, its mass by the sum of protons and neutrons. The number and distribution of electrons are the same in all of the isotopes of an element since here the nuclear charge plays the only role, whereas atomic masses differ.

This explanation might also support the idea that chemical properties are defined solely by the electron shell structure. However, more detailed investigations might raise some well-founded doubts regarding this argument. Exact measurements revealed that substances differing only in isotope compositions do behave differently. H_2 and D_2 or H_2O and D_2O, called "light water" and "heavy water", differ from each other and not only as far as freezing points, melting points and optical spectra are concerned. Traditionally, these properties were regarded as "physical", although it is difficult to justify any difference between "physical" and "chemical" properties in view of statistical physics and quantum mechanics. Nevertheless, the stabilities and rates of formation of compounds also change as one isotope is exchanged for another. Thus, it is not only the electron shell what counts!

Frankly speaking, these differences, called isotope effects, are small and only play some role with the lightest of elements where the relative mass differences are large. This shows indeed that the decisive role in chemistry is played by the electron shell. Its existence, however, shows that atomic masses also count. This is no particular wonder. Properties like gas pressure or specific heat were understood in terms of molecular vibrations. Therefore, atomic masses upon which vibration energies depend necessarily influence the behaviour of the substances.

Isotope effects are important in processes of the nature. For example, the proportion of heavy water to light water is different in sea and in springs. The rates of chemical reactions, evaporation and crystallization are also isotope dependent, some of them are faster with the lighter, others with the heavier isotopes. According to the processes involved, the isotope compositions of the products become different. This enables one to trace down the details of geological and meteorological processes of the past.

The existence of isotopes caused some trouble regarding the definition of the atomic mass unit. The nineteenth century unit based on the density of the hydrogen gas has been seen to be ill-defined because the isotopic composition of the gas changes with its origin. The unit must be based on a given isotope of an element. According to present-day agreement the atomic mass unit, AMU is equal to one-twelfth of the mass of an unbound atom of the carbon-12, the ^{12}C isotope. The choice of a carbon isotope has a practical reason: the nuclear masses of carbon isotopes can be measured the easiest and with good reproducibility through mass spectrometry. This definition makes the atomic masses differ but little from the H gas scale, to which chemists are traditionally accustomed. Expressed in unit of mass 1 AMU equals $1.66053904020 \times 10^{-27}$ kg.

17   Water

Talking about water, it is difficult to resist the temptation to quote Thales: *"Water is the first principle of everything."* Our present world view being much different from Thales' we know even better that terrestrial life had its origin in and depends on water, as it is shown by both the history of biological evolution and our everyday life, exposed as it is to the threats of floods and droughts. Hydrogen is the third most abundant element on Earth being

Fig. 42 Electrons among the atoms in the water molecule. A chemical bond is due to the electrons which simultaneously belong to two atomic cores

mostly present in the form of water, the total mass of which is as high as 2×10^{18} tons.

Atoms in the H_2O molecule are kept together by two covalent bonds. The readiness to form a rare gas shell, often called the "octet principle" is clear in that case. As can be seen in Fig. 38, oxygen has six electrons; the two hydrogens have one electron each. The electrons are distributed among the atoms as it is shown in Fig. 42, that is two pairs of electrons belong both to the oxygen and to one or the other hydrogen; thus, two electrons are around each hydrogen (He configuration) and eight around the oxygen (Ne configuration). A chemical bond means a pair of electrons that belong to both of the atoms. This is the model-like description of the "valence stroke", the structural formula H–O–H. The real essence of the covalent bond was explained through the example of the H_2 molecule.

As a first approach to the structure of any molecule, let us recall the "electron clouds", the electron density figures, of the hydrogen atom (Fig. 34). Electrons in the atom of any element are distributed approximately similarly to the possible electron states in hydrogen. Two of the eight electrons of O are of a distribution similar to the two lowest H states surrounding the nucleus as a spherical cloud, this is the He shell. Two of the further six are also of a spherical distribution but with a higher energy whereas the remaining four are expected to produce three, mutually perpendicular shells, like the next H shells; this expectation is given in Fig. 43. One of them is "saturated" i.e. two electrons of opposite spins are there, the remaining two housing one electron each thus being open for taking another electron of opposite spin. That is why oxygen has two valences.

Only the three, mutually perpendicular shells were drawn in Fig. 43. Looking at the figure, one would expect the two valences of O to be perpendicular to each other because valences correspond to the half-filled shells. Accordingly, the two O–H bonds in a water molecule should subtend at 90°. Experiments falsify this expectation.

X rays or electron beams are scattered on molecules similarly to a light beam being scattered on a grain of dust or, as it was discussed earlier, on sites of uneven densities. The short wavelength of the X ray is the reason why it is scattered by objects as small as a molecule. The change in the orientation of

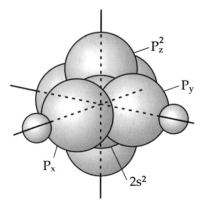

Fig. 43 Electron shells of a free O atom would be like that if they were completely similar to the H atom shells

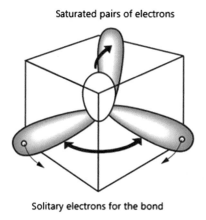

Fig. 44 Electron structure of the H_2O molecule

the wave propagation discloses the size of the scattering object, provided the wavelength is known. Once the distances O to H and H to H are known, the shape of the H–O–H molecule can be evaluated.

That direct measurement can be supported by other, less direct ones. Molecules of different shapes rotate differently, hence the part of optical spectrum which is due to rotation and also molar heat reflect the shape of the molecule. All these results show that the valence angle in the water molecule is much different from right angle it being 109.45°. The shape of the water molecules is given in Fig. 44. Was the counting of the electrons in error? No, the error was in the translation of the H atom states to other atoms without any alteration. One has to take into account the spherical shells the energies of which are almost equal with those of the perpendicular shells. Also they participate in the valence bond. If only the spherical shells produced the bond the molecule would be linear because the two H atoms repel each other due to the Pauli principle, i.e. to the incompatibility of identical state electrons. The observed geometry indicates a trade-off by the O atom: the chemical bond is brought about in one part by the spherical electron cloud, in another part by the two perpendicular shells.

Six electrons in the water molecule are very similar to each other. Two of them are parts of the valence bonds, whereas the four other take their places in pairs in two shells where the electron density is consequently very high. These are closed shells where no further electron can be accommodated. Indeed, a water molecule is observed to be chemically most stable.

Despite all that, some curious effects can be observed if the density of the water vapour increases or if the vapour is even liquefied. Elementary measurements make the impression as if the molar mass increased as vapour density increases. Since the times of Avogadro and Cannizzaro it has been well known that gas densities, taken at the same temperature and pressure, are proportional with the molar masses, at least as far as perfect gases are concerned. Gas density determinations indicate an increase of molar mass with vapour density. Even more direct measurements like mass spectrometry, although it is limited to low pressures, yield similar results. While molecular mass 18 appears in the mass spectrum, corresponding to the formula H_2O, also masses twofold and several fold of this value are present.

Despite lacking free valences water molecules can closely couple. The two electron pairs, playing no role in O–H bond formation, interact with the H atom of another water molecule. This is no real chemical bond since there is no way for new interatomic electron pairs to form. Still, an attraction exists between the electron pair of the one and the H nucleus of the other molecule. This interaction is called hydrogen bonding. Provided water density is high, more than two molecules might couple in that way bringing about the structure seen in Fig. 45. Water molecules form almost a crystal lattice, similar to ice structure. The kinetic energies of the molecules are high enough to overcome the attraction between the neighbours if temperature is above melting point. Hydrogen bonds are easily broken or deflected, thus vapour or liquid properties prevail. Still, the surprisingly high boiling point of water, it being much less volatile than H_2S although it is of similar structure and of lower molecular mass, clearly indicates that quite a fraction of hydrogen bonds remains untouched above melting point.

The close coupling of the molecules makes it difficult to tell which atom belongs to which molecule. Even more, some of the molecules might decompose because the products are firmly bound to the neighbouring molecules.

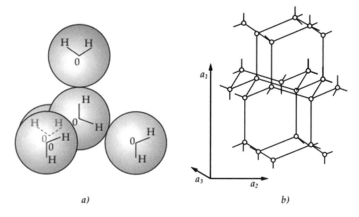

Fig. 45 Water molecules in condensed phases. **a** Molecules in liquid water, **b** ice crystal

An example of such decomposition, called water dissociation, is given below with the dotted lines symbolizing H bonds.

$$H_2O \cdots H_2O \cdots H{-}O{-}H \cdots OH_2 \rightleftharpoons H_2O \cdots H_3O^+ + {}^-OH \cdots OH_2 \tag{17.1}$$

The double arrow indicates that both directions of the transformation are possible. Hydroxonium ion, H_3O^+ and hydroxide ion, OH^- are bonded to neighbours strong enough to make the molecule decompose into ions, although the probability of decomposition is low.

Electric conductivity enables one to determine that probability which is found low indeed. 1 kg water contains only 10^{-7} mol in dissociated, ionized form at room temperature. In other words, the concentrations of H_3O^+ and OH^- ions are 10^{-7} mol/kg.

These two ions may also be of a different origin. If an acid, e.g. hydrochloric acid, HCl is added to pure water, two kinds of ions, Cl^- and H_3O^+ are released,

$$HCl + H_2O \rightleftharpoons H_3O^+ + Cl^- \tag{17.2}$$

There are more H_3O^+ ions in a hydrochloric acid solution than in pure water and this compels some of the ionized molecules to unite again forming H_2O. As the amount of H_3O^+ increases, that of OH^- decreases. Proceeding the other way round, the amount of OH^- ions can be increased by dissolving some alkali, e.g. sodium hydroxide, NaOH,

$$NaOH \rightleftharpoons Na^+ + OH^- \tag{17.3}$$

In that case the amount of H_3O^+ ions decreases.

The direction of the lower arrow becomes more pronounced in the equation of water dissociation if either acid or alkali is added to the water. This agrees with one's naïve expectation, since if there are numerous H_3O^+ ions around they find it easy to react with an OH^- ion. Some general reasoning will follow later.

The product of the concentrations of H_3O^+ and OH^- ions in any solution is constant changing only with temperature,

$$K_{water} = c_{H_3O^+} c_{OH^-} \tag{17.4}$$

Expressing concentrations by mole/kg, $K_{water} = 10^{-14}$ (mole/kg)^2 at 20 °C.

Let us be reminded that "logarithm of a given number is an exponent" and "the logarithm of a product is the sum of the logarithms of the factors" as it was taught at school. Chemists have a predilection for the use of logarithm. The above equation in logarithmic form reads as

$$\log c_{H_3O^+} + \log c_{OH^-} = \log K_{water} = -14 \tag{17.5}$$

The sum of the logarithms of the ion concentrations equals -14 in room temperature water. An aqueous solution is called neutral if the concentrations of the two ions are equal, $c_{H_3O^+} = c_{OH^-} = 10^{-7}$. Hence, the negative logarithm of the H_3O^+ ion concentration is 7,

$$-\log c_{H_3O^+}^{neutral} = 7 \tag{17.6}$$

As shown by experimental practice it is more difficult to measure H_3O^+ concentration than its logarithm. Thus it is expedient to give a separate name to the logarithm of the concentration it being called hydrogen exponent and denoted as pH. Its definition is

$$-\log c_{H_3O^+} = pH \tag{17.7}$$

Obviously, pH = 7 prevails in neutral water at room temperature. Adding acid to the solution, pH decreases, whereas adding alkali it increases.

Hydrogen bonding is no peculiarity limited to water molecules. Similar, admittedly weaker bonds form between ammonia, NH_3 hydrogen fluoride,

HF or hydrogen sulphide, H_2S molecules. Moreover, hydrogen bonds play an extremely important role in biology. The two strands of the double helix of deoxyribonucleic acid, DNA are held together by hydrogen bonds. This book is certainly not the right place to discuss the role of DNA in cell biology and heredity. Still, let us pause for a moment: was Thales right perhaps?

Further Reading

P. Atkins, *Physical Chemistry: A Very Short Introduction* (Oxford University, Oxford, 2014)

P. Atkins, J. de Paula, *Atkins' Physical Chemistry*, 8th edn. (W.H. Freeman, New York, 2006)

P.W. Atkins, *Physical Chemistry*, 4th edn. (Oxford University Press, Oxford, 1990)

L. Pauling, *The Nature of the Chemical Bond*, 3rd edn. (Cornell University, 1960)

S.G. Brush, *The Kind of Motion We Call Heat*, vols. 1, 2 (North Holland/Elsevier, Amsterdam, 1986)

Which Way and How Fast

1 Hydrogen and Fertilizer

Talking about ecology and nature conservation most people tend to think that all the malaise inflicted upon present-day humanity arose with the recent decades' rapid industrialization, the threatening multiplication of cars, perhaps the growth of densely packed sea-side resorts. There are some who try to protect the world against the unlimited use of deodorizing sprays. The danger our natural habitat is exposed to is indeed most serious and must be faced in a serious and determined way. It is, however, to be understood that all these troubles started much earlier than some half a century. From the earliest times, mankind has damaged its environment with all its rational activity, being the more harmful, the more it deserved the name of a human being. The hominids of the Olduvai gorge did less harm than the inhabitants of Niniveh, Norman sailors less than the thriving chemical and engineering industry of the South East of England. That is what homo sapiens is like. It is our pressing and immediate task to minimize this destruction in order to deserve the name of a human being even better.

There is but little doubt that it has been systematic agriculture which has exerted the cruellest influence on nature's undisturbed cycles. Without humans, the carbon dioxide and the nitrogen of the air are fixed by green plants and nitrifying bacteria, respectively, producing organic compounds built mainly of carbon, hydrogen, oxygen, and nitrogen. They are the bases of life. The animal world is fed by them, which transforms the substances according to its own needs. The fallen fruit and the animals' manure send back to the soil what was robbed by the growing plants. The farmer greatly disturbed that cycle by eating up the crop or giving it as fodder to the animals. The increasing population consumed more and more nutrient, particularly nitrogen containing substances, without caring or being able to provide the arable land with nitrogen. For millennia people were taught by experience that the land should be laid fallow for some time, animal manure is to be strayed over the soil or clover should regularly be planted, however, all these methods were unable to protect the shrinking plough-lands from the appetite of the burgeoning population.

At this point, the chemist had to step in. In the middle of the nineteenth century, Liebig suggested putting some inorganic substances into the impoverished soil, since plants can make use of them in building their own organisms. In short, he advised the use of fertilizers. Soon it became clear that there is a shortage of inorganic nitrogen compounds. The majority of terrestrial nitrogen is present in the atmosphere in elementary form, nitrogen containing minerals are few, saltpetre and Chile saltpetre, chemically speaking KNO_3 and $NaNO_3$, being the two most important of them. These are water soluble compounds; hence, they can be used without further chemical treatment. The happy land owners in Chile were made prosperous by this discovery, when their vast arid and sandy lands turned out to be rich in $NaNO_3$ sources. For decades, modernized agriculture depended on the Chile saltpetre mines.

The nitrogen content of terrestrial atmosphere is obviously an inexhaustible source. Gaseous nitrogen, however, reacts most sluggishly; nature herself seems to be unable to accelerate the chemical transformation. That is why nitrogen is kept as an elementary gas in the air. The chemical industry had to solve the task of nitrogen fixation in order to meet the demands of contemporary agriculture. A reaction was to be found and industrialized by which atmospheric nitrogen is transformed into a water soluble nitrogen compound, effective in plant life.

Finally, after a series of half-successes, of ammonia, NH_3 turned out to be the solution. This compound, more precisely one of its salts, ammonium chloride, NH_4Cl has been known for ages. Herodotus reported about blocks of that salt found in the Libyan desert where the cult of the god, Amon prevailed, hence the name.

The water soluble ammonia gas, of pungent odour, is not itself a fertilizer. It can be oxidized to form nitric acid the salts of which, even its ammonium salt, act as excellent fertilizers. The relevant reactions are easy to put down their industrial implementation, however, being a serious task. Ammonia is formed from N_2 and H_2,

$$2N_2 + 3H_2 \rightleftarrows 2NH_3 \tag{1.1}$$

Ammonia can be oxidized to nitric acid,

$$NH_3 + 2O_2 \rightarrow HNO_3 + H_2O \tag{1.2}$$

When dissolved in water it turns into ammonium hydroxide,

$$NH_3 + H_2O \rightleftarrows NH_4OH \tag{1.3}$$

Finally, nitric acid and alkaline ammonium hydroxide form a salt,

$$NH_4OH + HNO_3 \rightleftarrows NH_4NO_3 + H_2O \tag{1.4}$$

Double arrow means here, as before, that the reaction might proceed in either direction under ordinary conditions, as it is usually expressed, they are reversible.

Most challenging of the above reactions is the first one where nitrogen is to be compelled to react. Brute force did not help here. Instead, the optimum conditions for a given chemical reaction, the possibilities of its acceleration were considered, the expected amount of the product was estimated. The practical aim was clear: much and fast. Ammonia synthesis was the first of the processes in the chemical industry the optimal conditions of which were determined by detailed physical–chemical calculations and measurements.

The importance of the process is seen by its unique, inestimably precious service rendered to agriculture. There is no other single chemical product that is produced in an amount as high as that one. (This is true if the amounts are expressed in moles; if expressed in mass, sulphuric acid outdoes ammonia.) The 95% of global hydrogen production is used for ammonia synthesis.

2   The Double Arrow—Chemical Equilibrium

The double arrow in a reaction equation expresses the fact that it lies with the actual conditions which of the substances are reactants and which the products. NH_3 might form from N_2 and H_2 but NH_3 might also decompose into N_2 and H_2. Such reactions always tend towards an equilibrium state. Chemical equilibrium means that there exists a certain proportion of the concentrations at which, if reached, reactions do not proceed any further, irrespective of the absolute amounts of the substances present. The word equilibrium refers to a mechanical analogy: the equilibrium of a balance depends on the ratio of the masses and not on their absolute magnitudes.

Let N_2, H_2 and NH_3 be closed in a vessel at fixed pressure and temperature; the substances undergo a reaction according to the equation given in the previous chapter on fertilizers. If the concentration of ammonia is low, synthesis prevails according to the upper arrow, if it is high, decomposition takes place according to the lower one. Equilibrium concentrations change with temperature and pressure, but the total amount of the substances, i.e. the volume of the vessel plays no role.

That much might be known almost by instinct. The theoretician's task is to determine the equilibrium state and to explore how the equilibrium concentrations can be changed so as to increase ammonia production. Expressed in the chemists' parlance, the question is how to shift the equilibrium in the direction of the upper arrow.

Let us re-write the equation which connects energy, U, entropy, S, and work, L given in Chap. 9 "The science of possibilities", as

$$\delta U = \delta L + T \delta S \qquad (2.1)$$

This expression, referring to small variations in its present form, holds for equilibrium conditions. That limitation is most important. If the system is out of equilibrium, its properties start varying so as to reach equilibrium state.

As the system moves towards equilibrium there appears an extra contribution to entropy. The simplest example of that most general law is that of a gas closed in a cylinder with a moving piston: should the pressure of the gas be increased or decreased by the piston, its friction produces heat whichever way it moves. That heat is accounted for by the extra entropy.

Limiting the present treatment to constant temperature and constant volume conditions, hence any mechanical work being impossible, $\delta L = 0$, the above equation writes

$$\delta U - T \delta S \leq 0 \qquad (2.2)$$

equality referring to equilibrium, inequality to non-equilibrium conditions.

Here lies the answer to the present question. Chemical reactions usually do not proceed under isolated conditions that keep the system's energy constant, but under isothermal conditions that keep temperature constant.

That involves a heat exchange between reaction vessel and its environment so as to maintain the temperature constant despite the heat effects of the reaction.

It is expedient to define a function, called free energy, A, as

$$A = U - TS \qquad (2.3)$$

The above inequality referring to δU and δS can be expressed as

$$\delta A \leq 0 \qquad (2.4)$$

meaning that A decreases during any constant volume, constant temperature process and does not change any more as equilibrium is reached. In the next passages we deal with the equilibrium situation. At equilibrium or near to it the expression $\delta A = A(N_2) + A(H_2) - A(NH_3)$ prevails.

There are two features of A that deserve attention. One of them is almost obvious: free energy is always proportional to the amount of the substances similar to volume or energy. The other one is most fortuitous: free energy is the sum of terms one part of which contains the properties of the pure compounds and another part which depends on concentrations. The equilibrium situation writes as

$$A_0(\text{pure substances}) + A_e(\text{mixture}) = 0 \qquad (2.5)$$

with the subscript e referring to equilibrium.

The main point of this expression is that the free energies of the pure substances, being known through calorimetric studies, define the extent of the mixture contribution. Beyond that it is well known how this contribution depends on equilibrium concentrations (this explicit equation not being included here). With all this information at hand, equilibrium concentrations can be evaluated by making use of the thermal properties of the pure substances. Performing all the calculations, one finds a much used expression,

$$K = \frac{[NH_3]_e^2}{[H_2]_e^3 [N_2]_e^2} \qquad (2.6)$$

The square bracket denotes the concentration of the bracketed substance with subscript e indicating equilibrium. The exponents are equal to the numbers of moles figuring in the reaction equation. The main thing is that K, called equilibrium constant, is independent of concentrations although it varies with temperature and pressure. It is expressed solely by the properties of the pure compounds.

It goes without saying that all these are true beyond ammonia production. Let a reaction be taken with the general equation

$$b\text{B} + c\text{C} \rightleftharpoons l\text{L} + m\text{M} \tag{2.7}$$

where capitals denote compounds, lower case letters the numbers of reacting moles. Equilibrium constant writes as

$$K = \frac{[\text{L}]_e^l [\text{M}]_e^m}{[\text{B}]_e^b [\text{C}]_e^c} \tag{2.8}$$

The rule is clear, so is the use of the expression. By knowing the initial concentrations and the value of K, both the direction and the endpoint of the process can be predicted. Components with concentrations higher than their equilibrium values must disappear, those with concentrations lower than those must form, until the above fraction, composed of actual concentrations, is equal to K. At that point the reaction stops.

Now the practical question regarding the increase of ammonia production can be answered. The mixture must be influenced so as to result in a high $[\text{NH}_3]_e$ value. Ammonia appearing in the numerator of the fraction, K has to be increased for that. As it was said above, K depends on the properties of the pure components. According to experimental determinations, the joint energy of three moles of H_2 plus 2 mol of N_2 is higher than the energy of 2 mol of NH_3. This means that as ammonia is formed from its elements heat is released. Such processes are called exothermic. The colder the system is the easier the evolved heat can escape. Or, to be more exact and less pictorial, the lower T is, the larger the entropy change becomes. Generally speaking, low temperature is advantageous for exothermic reactions: equilibrium concentrations of the products are higher at lower temperatures.

Gaseous reactants are transformed into a gaseous product as ammonia is synthetized, 5 mol of reactants being transformed to 2 mol of products. That decrease of the number of moles implies the decrease of the total volume of the gaseous mixture. At constant pressure the volume deceases as ammonia is being formed. Hence, a decrease in volume, i.e. an increase in pressure, promotes ammonia formation.

Ammonia production is exothermic, also accompanied by volume decrease. Low temperature and high pressure are advantageous to such reactions. The opposite is also true; if heat is adsorbed during a reaction, such processes being called endothermic, high temperature is beneficial. If a reaction is accompanied by volume increase pressure should be kept low.

The general principle of *Le Chatelier* and *Brown* is often quoted in this context. It states that any system exposed to some external effect behaves so as to weaken the result of the effect. That is, an increase of the external pressure helps processes where the volume decreases, and cooling enhances exothermic reactions.

3   Time Counts—Rates of the Changes

So far, we have not touched the obvious question of how long processes take. Writing about heat exchange, we did not ask how fast heat goes over from one system to the other; talking about vaporization and melting, mixing and combination no mention was made of the period one has to wait until liquid turns into vapour, crystal into liquid, nitrogen into ammonia. At the present point, however, we are compelled by economy to consider that problem. The costs of some industrial chemical process is set not only by the amounts of products, reactants and the energy needed. Costs are influenced also by the rates of the reactions. The faster the reaction is the smaller or the fewer reactors are needed for the production of a given quantity. Here investment and operational costs, which are considerable in modern chemical industry, have an important say.

It is not only money that compels us to deal with process rates. The details of a process cannot be understood if only its equilibrium conditions are known. Quoting a mechanical analogy again, the causes of motion cannot be studied through the rules of equilibrium. Gravitational field would be impossible to describe by making use of a balance at equilibrium. That would only show that gravitation acts equally upon the two pans. The nature of gravitation can be revealed by observing a pendulum in motion, a stone falling or the planets orbiting.

Both practical requirements and theoretical curiosity take us to the investigation of the kinetics of processes in physical chemistry. Therefore, it is surprising why we did not start with kinetics. Because equilibria are easier to study or because that is the historical route? No, these are evasive answers. It is better to admit that equilibrium thermodynamics is, even today, superior to kinetics as far as conceptual clarity and theoretical foundations are concerned.

There exist general statements regarding kinetic processes; their practical values, however, are sometimes equivocal. According to the Second Law, the entropy of a closed system increases in any natural process, if $\delta S > 0$ prevails. The increase of entropy during unit time, called entropy production, is positive until equilibrium is attained—at which point it becomes zero,

$$\frac{dS}{dt} \geq 0 \tag{3.1}$$

[The rate of entropy change is written here as a differential quotient, similarly to the expression of velocities in mechanics. The two equations have similar meanings. S does not always increase by the same extent during equal time intervals; hence, the rate quotient $(S_2 - S_1)/(t_2 - t_1)$ is meaningful only at the $(t_2 - t_1) \rightarrow 0$ limit. The idea of the differential quotient will not be explained any more.] The higher the entropy production of a process is, the faster the system approaches its equilibrium.

There exist theoretical methods to evaluate the entropy production of certain processes in terms of experimental parameters, like amounts of substances, temperature, volume, pressure. Let us recall that entropy cannot be measured! These theories, however, seem to be more important in summarizing earlier experiences than in predicting new laws. The rates of changes of system properties like temperature, the amount of substances, their composition, etc. are mainly described by empirical, observation-based relationships.

One of the earliest laws of this kind is that of heat conduction, expressing the amount of heat which flows from the hotter to the colder spot in a body.

The simplest version of the problem is given in Fig. 1. Body temperature is different from point to point at x_1 being T_1, at x_i being T_i. The law states that the amount of heat that flows during unit time is proportional to the cross section of the body, F and—most important!—with the temperature

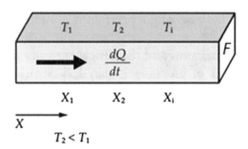

Fig. 1 Heat conduction

difference per unit length, dT/dx. Mathematically,

$$\frac{dQ}{dT} = -\kappa F \frac{dT}{dx} \tag{3.2}$$

The negative sign expresses that heat flows towards lower temperature sites, this being the requirement of the Second Law.

A similar empirical law refers to diffusion, the process which makes the concentrations uniform without any mixing or stirring. The simplest way to produce an uneven concentration distribution (Fig. 2) is to cautiously let flow a small amount of concentrated solution from a pipette underneath pure water. If copper sulphate solution happened to be chosen, the blue colour makes it easy to follow the spread of the substance with the naked eye.

Going upwards in the vessel concentration decreases, that being a function of the position, $c = c(x)$. Similarly to heat conduction, the amount of substance that flows during unit time, dn/dt, is proportional to the cross section of the vessel, F and with the concentration difference per unit length, dc/dx,

$$\frac{dn}{dt} = -DF \frac{dc}{dx} \tag{3.3}$$

The negative sign expresses that the dissolved substance wanders towards sites of lower concentrations. This reflects the Second Law again, the entropy being highest in the homogeneous solution where c does not depend on x.

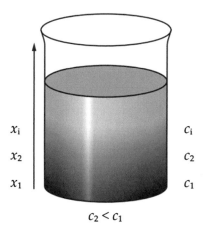

$c_2 < c_1$

Fig. 2 Diffusion

The rate of a chemical reaction is expressed by the concentration change per unit time. Ammonia production is fast if the ammonia proportion increases or, what amounts to the same thing, those of hydrogen and nitrogen decrease quickly. It goes without saying that reaction rate is also defined as a differential quotient,

$$\text{reaction rate} = \frac{dc}{dt} \tag{3.4}$$

Unfortunately, a simple general law similar to heat conduction or diffusion, is unknown for chemical reactions. Small wonder in view of the great variations of chemical reactions. The sign of rate tells us only that it is positive for the products and negative for the reactants, $dc_{product}/dt > 0$, $dc_{reactant}/dt < 0$.

It is often found that the reaction rates are proportional to a power of the concentrations. The simplest of cases is when the rate depends on the first power of only one concentration, rate and concentration being proportional to each other. An example is the decomposition of N_2O_4, a substance figuring in nitric acid manufacturing, according to the reaction

$$N_2O_4 \rightarrow 2NO_2 \tag{3.5}$$

The rate of decrease of N_2O_4 concentration is proportional to its actual concentration,

$$\frac{d[N_2O_4]}{dt} = -k_N[N_2O_4] \tag{3.6}$$

square brackets denoting concentrations again but now with the subscript e missing. The system is out of equilibrium as far as reaction goes on.

The reaction between hydrogen and iodine vapour results in hydrogen iodide, HI, a compound akin to HCl. Production rate is proportional to the concentrations of both of the reactants,

$$\frac{d[HI]}{dt} = k_{HI}[H_2][I_2] \tag{3.7}$$

This equation is seductive in its simplicity. It suggests that hydrogen iodide is being produced fast if hydrogen and iodine molecules meet with a high probability. Sure, because concentration means the amount of substance in unit volume of the mixture or, in other words, it is the probability of finding a molecule in a space of unit volume. The product of concentrations is the probability of two molecules to be in the same unit volume. (Multiplication of probabilities, we have already discussed in the context of thermodynamic probability and entropy.) It seems, two molecules react if they meet, provided they are prone to undergo some chemical transformation. A high probability of encounter amounts to a fast reaction.

Is this the gist of reaction kinetics, the science of chemical rates? Perhaps not quite. It is not for nothing that coefficients, k in the rate equations are denoted by different subscripts, indicating that they are different for each reaction. The rate coefficients, as these parameters are called, do not depend on concentrations but very much on temperature. Apart from rare exceptions, higher temperatures mean higher rate coefficients, hence faster reactions.

The average energy of the molecules is proportional to temperature; hence, k increasing with T might indicate that molecular encounter is not a sufficient condition of a reaction taking place, and that some extra energy is needed. Higher average energy results in faster chemical transformation. This looks quite natural. Both H_2 and I_2 are stable molecules each being held together by strong covalent bonds. For any reaction to proceed the bonds must be loosened first to make the atoms able to form new bonds in a new molecular arrangement. Only those molecules are ready for a transformation at the moment of encounter which are so energetic as to break the old valence bonds. Energies of the initial and final states do not decide alone the fates of the molecules. Although two molecules of HI are energetically more favourable than one molecule of H_2 and I_2 together, something more is needed for the reaction. The encountering reactants must have the necessary energy to open their bonds.

An outline of the situation is given in Fig. 3. The joint energy of two HI molecules is lower than that of one H_2 and one I_2 molecule. The reactants, however, have to overcome an energy barrier which is higher than either

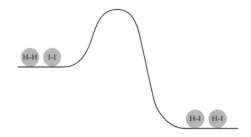

Fig. 3 Reactants must overcome a potential barrier to complete the reaction

the initial or the final energy level. In short, reactants must be activated to make the reaction proceed. The process of activation is expressed by the rate coefficient the necessary energy being called activation energy.

The energy of each molecule is different. Those molecules are able to react whose energies are higher than the potential barrier. As temperature increases the high energy fraction of the molecules also increases, that is the explanation of the rate coefficient's dependence on temperature. If activation energy is high, the rate coefficient depends markedly on temperature.

Now, is it true that whereas the product of concentrations expresses the frequency of molecular encounters, rate coefficient accounts for the energy needed for activation? Well, almost. Detailed studies of the reaction rates require a more subtle explanation.

Comparing the response of the reaction rates to changes in temperature a good number of reactions can be found that behave similarly, showing that their energies of activation do not differ markedly. Still, their rate coefficients are much different. It seems, k expresses something more than the height of the energy barrier.

Molecules, simple as they might be, are not spheres without any structure; they have definite shapes and various volumes. Also, there are different kinds of motion: translation, rotation, bond oscillations proceed simultaneously, the total energy being distributed over these possibilities. The encountering molecules must obtain a certain relative position for the reaction to take place. Only few of the relative positions given in Fig. 4 and a number of similar ones enable the molecules to react.

Moreover, even if both total energies are sufficient and relative positions advantageous the reaction is still not certain to proceed. The energy must be amassed at that part of the molecule where it is needed. H_2 and I_2 react if the H–H and I–I bonds vibrate energetically. Fast translation or quick rotation does not help. A particular valence bond must have high energy.

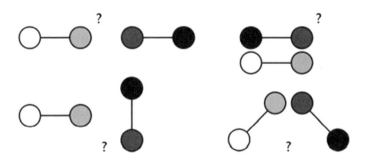

Fig. 4 A number of relative positions in reacting molecules

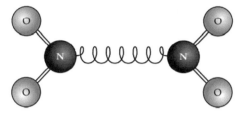

Fig. 5 The vibration of an N_2O_4 molecule necessary for decomposition

To summarize what was described above we may write,

reaction rate $= k \cdot$ encounter probability

k : sufficient energy at the right place $+$ appropriate geometry

If we understand the basics, reactions like the decomposition of N_2O_4 looks problematic. Why does a molecule decompose of its own without colliding with other molecules? If it is unstable why does it not decompose immediately? If it is stable what makes it decompose after a period of time? Although a single N_2O_4 molecule is the sole source of the two NO_2 molecules there is a need of the assistance of the neighbouring molecules. They must deliver the necessary energy to the molecule to be decomposed, activating it for the reaction. It must overcome a potential barrier similar to the one in Fig. 3 and then the energy must be amassed on the valence bond to be cleft (Fig. 5). The important difference from the HI reaction is that no other molecule is involved in the reaction; the neighbours deliver only energy. This is expressed by the absence of any other concentration but that of the N_2O_4 in the rate equation.

The proportionality between rate and concentration expresses only the fact that if there is more reactant per volume the process is faster.

4   Real Processes Are Rarely Simple

There are but few processes the kinetics of which would obey the simple rules described above. Rates of chemical reactions are not always proportional to concentrations and the mathematical relationships are often involved. The rate is sometimes influenced by substances which seemingly do not participate in the process. Let us take the oxidation of SO_2 as an example,

$$2SO_2 + O_2 \rightarrow SO_3. \tag{4.1}$$

Still, if nitrogen oxide, NO is added to the gas mixture the process gets much faster although the reaction equation does not refer to that compound.

The high number of reactions defying simple rules makes one wonder whether the model and theory described above, despite their firm physical basis, are correct. Only direct experimental proofs can dispel one's doubts.

The real touchstone of the molecular models is the investigation of the fate of individual molecules instead of the behaviour of a multitude. The encounters, collisions, motion, energetics of single molecules are to be observed, even influenced in order to compare model and reality. Statistical laws should be underpinned by observations of individual particles.

Nowadays, there are a stock of highly developed methods and instruments built for that purpose. Experiments are not limited anymore to molecular assemblies consisting of particles of different energies that obey Boltzmann distribution. Atoms, ions or molecules of sharply defined energies can be produced. (The idea of a velocity sieve was mentioned earlier in the context of the Maxwell demon.) These particles of known and well-defined energies meet and react in a small space. Reaction products leave the space their directions, velocities, masses, and energies being measured by instruments similar to a mass spectrometer.

The outline of such an apparatus is given in Fig. 6. The reagents in the form of perpendicular beams of pre-defined (sieved) velocities reach the reaction space. The products of different masses and velocities leave the space. The reacting pairs are studied almost one by one.

The study of the oncoming reaction partners and of the products support the previously described model of the chemical reactions. It has been shown that whereas encounter probabilities, energies, relative molecular positions

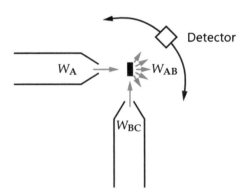

Fig. 6 Outline of a crossed beam apparatus for chemical kinetic studies; the reaction A + BC → AB + C reaction is investigated, the velocities, W of A and BC being set and that of AB observed

differ for each process, these factors together define the probability of the macroscopically observable transformation. In several cases, one can even pinpoint a particular valence vibration or electron state the excitation of which enhances the reaction. This can be done by illuminating the beams since excitation means the adsorption of light of a given frequency. Molecules of definite energy states can be produced by illuminating the beams with intense light, probably of a laser.

The model of pairwise collisions turned out to be correct. If so, there is a want of explanation for a good number of reactions where macroscopically determined reaction rates are not proportional to the product of the reactant concentrations. For example, in contrast to the HI case, the formation of two other hydrogen-halogen compounds, HCl and HBr, do not comply with this simple expectation.

Most of the reactions do not proceed in one single step but along a series of elementary stages each of which consists of pairwise encounters. Simple rules given above for the examples of HI formation or N_2O_4 decomposition hold for those stages. Macroscopic measurements, however, account only for the final outcome of the series of elementary stages; the overall process often depends on time in an intricate way. The analysis of the elementary stages is no easy business. A great chemist of the past scornfully called the endeavour to describe them: "paper chemistry".

Let us consider, for example, HI formation, if a good amount of HI is also present in the gas mixture. In that case, the decomposition of HI must also be accounted for,

$$HI + HI \rightarrow H_2 + I_2 \qquad (4.2)$$

In that case the reaction rate obeys a more complicated law,

$$\frac{d[HI]}{dt} = k_{HI}^{+}[H_2][I_2] - k_{HI}^{-}[HI]^2 \qquad (4.3)$$

the first term on the right hand side describing formation, the second one decomposition. Each of the terms complying with the simple expectations the two together result in a more complicated dependence on time. The expression is supported by the measurements.

The classical task of reaction kinetics is to explore the elementary stages of a macroscopic reaction. Two types of elementary stages are usually considered. The rate of one of them is proportional to the concentration of a single reactant, the other to the product of two concentrations. The first type is called first order, the other second order reaction. More complicated elementary reactions occur most rarely.

For example, hydrochloric acid forms from hydrogen and chlorine in a way that is very different to hydrogen iodide formation. Kinetic data, i.e. the dependence of rate on concentration and the effect of temperature on the rate coefficient indicate that what is observed is the outcome of a number of consecutive elementary stages. First, a Cl_2 molecule decomposes into two $Cl\cdot$ atoms, the dot symbolizing the unpaired electron, i.e. the "free valence",

$$Cl_2 \rightarrow 2Cl\cdot \tag{4.4}$$

Free chlorine atoms react with H_2 molecules,

$$Cl\cdot + H_2 \rightarrow HCl + H\cdot \tag{4.5}$$

The free $H\cdot$ atom attacks a chlorine molecule,

$$H\cdot + Cl_2 \rightarrow HCl + Cl\cdot \tag{4.6}$$

this series going on and on until two hydrogen atoms or two chlorine atoms meet to terminate the process,

$$Cl\cdot + Cl\cdot \rightarrow Cl_2 \tag{4.6}$$

$$H\cdot + H\cdot \rightarrow H_2 \tag{4.7}$$

All the elementary steps are first or second order processes, whereas their outcome is described by a complicated expression.

Similar processes, called chain reactions, are more or less common. They have an appeal to practical experts in view of their high rates and to theoreticians because of their intriguing complexity.

5   How to Produce Ammonia

If you want to produce ammonia from H_2 and N_2 you face an obvious contradiction. As ammonia forms, heat develops; it is an exothermic process. Having discussed the equilibrium of this process and of similar ones in a previous chapter, we could show that low temperature is advantageous: the cooler the gas mixture is the more ammonia is present as the reaction reaches completion. Reaction rate at low temperature, however, is low. But you wish to produce much and fast, because of economic reasons. To find conditions at which the formation reaction is both effective and rapid seems to be almost impossible. The global ammonia production being some 240 million tons per year disproves this concern, however.

As to pressure the situation is more favourable. If pressure gets higher both the equilibrium concentration of ammonia and its rate of formation increase. Whereas equilibrium is shifted in accordance with the Le Chatelier-Brown principle discussed earlier, the rate also increases, because if the pressure is high the amounts of substances per volume, i.e. the concentrations, are high.

You have to find conditions which have room for a compromise. On the one hand, you have to circumvent the low output at high temperature with some clever technical trick; on the other hand, you have to find a reaction sequence whose steps do not depend very strongly on temperature. With low energies of activation, a reasonable rate is possible even at modest temperatures.

That last requirement may look odd. It points to the possibility that whereas the initial substances, hydrogen and nitrogen, are given, the desired product may be obtained along more than one reaction path. That is perhaps not too amazing. A number of elementary reactions can result in one and the same product. An expedient choice must be made.

The basic idea of the practical ammonia synthesis is that the reaction is not performed in the gas space. It takes place on the surface of a solid substance. Technically speaking it is not a homogeneous process where all the reactants and products are in one and the same phase but it is a heterogeneous one where the substances are in different phases; part of them in the gas space, another part on the solid surface. The synthesis takes place on a well-prepared surface of elementary iron. N_2 molecules get stuck at the surface this process being called adsorption,

$$Fe_{solid} + N_{2gas} \rightarrow (Fe \cdot N_2)_{surface} \tag{5.1}$$

Hydrogen molecules of the gas phase hit the solid and react with the adsorbed nitrogen molecule giving ammonia,

$$(Fe \cdot N_2)_{surface} + 3H_2 \rightarrow (Fe \cdot 2NH_3)_{surface} \tag{5.2}$$

Finally, ammonia molecules leave the surface, that is, they get desorbed,

$$(Fe \cdot 2NH_3)_{surface} \rightarrow Fe_{solid} + 2NH_3 \qquad (5.3)$$

Obviously, the overall rate of the reaction sequence is determined by the slowest of the steps. The slowest step has the highest activation energy; thus, it is not just slow but its rate depends on temperature the most. This reaction is nitrogen adsorption determining synthesis temperature to be about 400 °C. Optimum pressure is about 200 bar. N_2 and H_2 have a very meagre inclination to react in the homogeneous gas phase under such conditions.

When the reaction is completed, the iron, although having played an important role in the synthesis, is found without any change either in form or in quantity. A substance that plays such a role is called a catalyst. Seemingly, it does not take part in the process, since its quantity neither increases nor decreases; it still enhances a reaction step. In reality, of course, it does react, in the above example it adsorbs nitrogen molecules. At a later stage, however, it is re-formed, assuming its initial form, as it was seen to happen at the last stage of the above process: ammonia gets desorbed; hence, iron is present again as it was at the beginning.

The iron catalyst makes the reaction faster but it does not influence the amount of ammonia formed. The equilibrium constant strictly predicts the percentage of ammonia in the gas mixture at the conditions of the reaction. One cannot outwit thermodynamics, only 15% ammonia is present by the end of the synthesis. The production and purification of nitrogen and hydrogen being expensive it would be a waste of money to squander 85% of the reactants. Instead, ammonia being easy to liquefy, it is frozen out of the gas mixture and the rest with the unaltered composition of reactants are fed into the reactor.

Atmosphere is an unlimited source of elementary nitrogen. Hydrogen, however, cannot be found naturally in its elementary form, it must be produced from water. One of the possibilities is to blow water vapour over glowing coal at about 1000 °C,

$$C + H_2O \rightarrow CO + H_2 \qquad (5.4)$$

Similarly, one can use methane instead of coal,

$$CH_4 + H_2O \rightarrow CO + 3H_2 \qquad (5.5)$$

Both reactions are heat absorbing, endothermic processes, thus either coal or methane must be heated first in order to make the reaction proceed. To that end, part of the coal or methane is burnt in air, also obtaining pure nitrogen, e.g.

$$2C + (O_2 + N_2)_{air} \rightarrow 2CO + N_2 \tag{5.6}$$

and only once coal reached the desired temperature is water vapour blown over the hot material to get hydrogen. This endothermic process cools the coal; thus, after a while, air is conducted to the coal again.

Periodically varying coal burning and water decomposition different compositions of gas mixtures are obtained; at one stage the mixture is rich in nitrogen, at the next stage it is rich in hydrogen. Removing carbon monoxide, CO, the mixture $N_2 + 3H_2$ can be obtained, which is needed for the synthesis to start with.

For more than a century, the chemical industry has been based on laws of thermodynamics and observations in chemical kinetics.

Further Reading

C. Capellos, B.H.J. Bielski, *Kinetic Systems* (Wiley-Interscience, New York, 1972)

K.J. Laidler, *Chemical Kinetics*, 3rd edn. (Prentice Hall, Hoboken, 1997)

J.W. Moore, R.G. Pearson, *Kinetics and Mechanism*, 3rd edn. (Wiley, New York, 1981)

Chemistry of the Outer Space

One would not expect any chemical reaction to take place in the outer space far from planets and stars. Two things are needed for a chemical phenomenon: matter and energy. The outer space contains very little of both, almost none at all. Gas is so dilute as to be most difficult to detect and energy comes only from star light and cosmic radiation.

Still, a good number of observations, several made before the start of space travels, indicate that the outer space is not a boring place for a chemist. The "Promethean feat" of Bunsen and Kirchhoff was already mentioned earlier. They applied spectroscopy to the light emitted by stars enabling them to determine their elementary compositions. This branch and technique of science is called emission spectroscopy, the investigation of emitted light. Its counterpart is absorption spectroscopy, which is based on the fact that cold materials, being unable to emit, absorb light. Each compound or element absorbs light at different wavelengths.

Putting a substance as a sample between the prism that decomposes light into its components of different wavelengths and a device that measures light, intensities at certain wavelengths are found to become weak (Fig. 1). They are absorbed by the sample. Any substance, whether oxygen of the air, the violet solution of potassium permanganate or a silicon crystal, has a characteristic "absorption spectrum", as the series of light intensities diminished by the sample at different wavelengths are called.

Absorption spectra are characteristic of the substances similar to emission spectra. Emission spectrum, however, can be observed only of those

R. Schiller, *A Non-Traditional Guide to Physical Chemistry*, https://doi.org/10.1007/978-3-031-07488-2_4

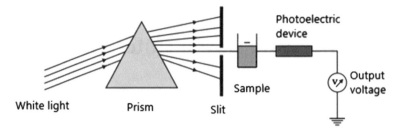

Fig. 1 Outline of an absorption spectrometer

substances which are stable at temperatures of incandescence. There is no such limitation with absorption, even the most unstable molecules have absorption spectra. The two processes are each other's mirror images. At emission the atoms release their energy in the form of electromagnetic radiation whereas at absorption they gain energy from the light source. Both processes obey the same laws of quantum mechanics, but only those light quanta are emitted or absorbed that correspond to the energy level differences of the atoms or molecules.

There are no emitting substances in the cold interstellar space. Only stars emit radiation. Light absorbing material, however, was observed in space. The first observed spectra were attributed to Na atoms and Ca^+ ions, later the number of substances increased and it is still increasing nowadays approaching two hundred. Chemists must have felt surprised that large molecules of complicated structures were also found. Starting with H_2, inorganic, compounds like water, ammonia, hydrochloric acid or NaCl and a large selection of organic compounds: acetylene, acetone, benzene, formic acid, even urea and glycine together with molecules as large as C_{60} and C_{70} (called fullerenes) were identified.

Some of them are also stable under terrestrial conditions, whereas others cannot be prepared or stored in laboratories. For example, both stable hydrogen cyanide, H–C≡N and its unstable isomer, H–N≡C were observed. The unstable compound was identified partly through theoretical calculations, partly by its preparation as a transient in electric discharge.

The density of the interstellar substance is low, not more than several particles per cubic centimetre. This is a trillion times smaller than the air density produced by a good vacuum pump. Density, however, might vary, since substances are able to form "clouds". Density in dilute clouds is about 1000 particles per cubic centimetre, whereas in dense ones it can reach 10 million. This is still very low compared to terrestrial conditions, corresponding to a pressure of 10^{-13} bar. Clouds also contain dust particles, their presence is shown by scattered light, similarly to what happens in a sun-lit room.

Incorrect, too light   Incorrect, too dark   Correct

Fig. 2 The glowing filament of the pyrometer must become invisible in front of the opening of the furnace

So there is little material in the space. What about temperature? It is 2.7 K, being equal to −270.5 °C; the answer seems to be much too quick and definite, an explanation is needed. In everyday parlance, temperature is something that is measured with a thermometer. The most obvious example is a mercury column the length of which is taken to be proportional to temperature. The concept here is that mercury attains thermal equilibrium with its environment and takes a volume determined by temperature (let us recall the notion of state equation discussed in connection with real gases). In fact, the thermometer measures its own temperature. If there is a high number of molecules in its environment, which exchange their energies with the mercury atoms, temperatures of environment and mercury are equal. Outer space, however, contains few atoms. A room temperature thermometer on the covering of a satellite cannot measure "the cold of the outer space".

Some thermometers work by a different concept. No thermometer can be housed in a high temperature furnace, it would be melted and evaporated at once. An optical pyrometer of Fig. 2 indicates the temperature by observing the light leaving the glowing furnace through an opening. Every glowing black body, or more exactly, every cavity surrounded by glowing walls, emits light. One can tell even with the naked eye that red colour means lower temperature and as the walls get hotter the light turns yellow and finally white. Industrial pyrometers do not analyze light, instead a filament is electrically heated until its temperature is equal with that of the furnace and the filament becomes invisible. The electrical current is calibrated against temperature.

More rigorous information can be obtained by making use of a prism and measuring light intensity. Figure 3 gives emitted light intensity at different wavelengths, changing with temperature. Two aspects are obvious at first sight: the wavelength of the highest intensity light becomes shorter and total light intensity (the area under the curve) becomes higher as temperature increases. Nothing else counts, emitted light being independent of the wall

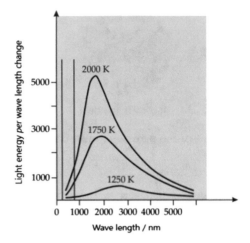

Fig. 3 Light emitted by glowing black body as a function of wavelengths at different temperatures

material, provided it is black. Black body radiation is determined solely by the body's temperature.

These curves played a most important role in the history of physics. By investigating the underlying laws Planck was induced to realize the quantized nature of energy, in other words, that the least amount of energy in a light wave of frequency, ω equals $\hbar\omega$. This was already mentioned in Chap. 12 about new physics.

Black body radiation is used for the measurement of the temperature of the space. Obviously, knowing that the place must be pretty cool, one had to count on electromagnetic radiation of low energy, that is, of large wavelength. Visible light is certainly not emitted by the interstellar space. However, microwave radiation of frequencies near that of radars delivers information from the space. While not emitted by stars or galaxies (such radiation also exists but it is not our present subject), it is a uniform background radiation, unrelated to heavenly bodies, streaming from every direction, filling the whole universe. It is understood as the black body radiation of the outer space. As if it were a furnace, cold as it might be, background radiation delivers information about its temperature. The task was to compare the observed spectrum with those similar to the ones in Fig. 3. It is all the same whether it is about visible light or radio waves. The comparison resulted in the temperature of the space.

Given the phrase about "the infinite cold of the outer space" this temperature seems to be high, one would expect 0 K by instinct. Refraining from cobbling in cosmology, we only dare quote the generally accepted views: background radiation is the evidence of the Big Bang as the origin of the universe. According to that theory, the primordial universe was of extremely small volume and of infinitely high temperature. Its sudden expansion resulted in our present world. Expansion, cooling, the formation of the simplest atom, H, later that of heavier atoms, all are the outcome of the Big Bang. Background radiation reflects the present stage of cooling. Gas clouds are much warmer than that their temperatures varying between 20 and 100 K. They are heated through the absorption of cosmic radiation.

In accordance with that theory, hydrogen is the most widespread element of the universe, amounting to about 90% of all the atoms, followed by helium that accounts for some 10%. The joint contribution of oxygen, carbon and nitrogen is less than 0.1%, that of sulphur, silicon, magnesium, iron and neon less than 0.01%. The rest of the elements contribute even less.

These are the conditions in the cosmic alembic where the compounds are being formed. Here even the presence of molecular hydrogen, H_2 is not obvious. A reaction can be written as

$$H + H \rightarrow H_2 \tag{1}$$

it is the route, however, along which hydrogen molecule would never form. The energy of the molecule is less than the joint energies of the two atoms, hence energy is released in the above reaction. If the molecule is unable to get rid of that it falls immediately apart. As if the atoms underwent an elastic collision. As if? This is an elastic collision in the strict sense, since a collision is inelastic if the kinetic energy of the colliding particles transforms into some different form of energy. This is impossible in the collision of two atoms. How is it then that we never observe free H atoms under terrestrial conditions because they all form molecules? That is because three particles always come together in our atmosphere,

$$H + H + M \rightarrow H_2 + M \tag{2}$$

and the third partner which might be a further H atom, an alien molecule or the wall of the vessel absorbs the excess energy. Here the two H atoms undergo an inelastic collision.

Material densities being very low in interstellar clouds three-body collisions are most improbable. It is impossible to account for the observed amount of H_2 by that mechanism. The formation of molecules is rather attributed to some catalysis-like process with interstellar grains of dust acting as catalysts.

At first, one H atom adsorbs on the surface of the grain, which is followed by a second one. These two walking at random on the surface, sooner or later they come near each other and unite giving the excess energy to the grain. Finally, the H_2 molecule desorbs, leaving the surface.

Although this process is very efficient, hydrogen is not present exclusively in the form of molecules. Light or cosmic radiation might decompose the molecules, i.e. they might undergo photolysis or radiolysis. The proportion of synthesis to decomposition is determined by gas density. If there are less than one hundred particles per cubic centimetre the cloud consists mainly of H atoms, if there are more than several hundred particles H_2 molecules prevail.

Most probably, molecules heavier than H_2 do not form along a similar catalytic route. Heavy atoms would move slowly on the surface and product molecules would desorb with difficulty. In spite of what was written above, heavier molecules form in two-body collisions, their excess energy being lost by light emission. Improbable as such a process might be, it must be taken into account in view of the practical impossibility of three-body encounters. For example, a CH_2^+ ion, unstable under terrestrial conditions, might form in that way,

$$C^+ + H_2 \rightarrow CH_2^+ + \hbar\omega \qquad (3)$$

ω meaning the frequency of the stabilizing photon. One should also consider that this ion contains two valence bonds contrary to the single one in H_2 thus excess energy is divided between two bonds. Half of the energy being insufficient for the scission of the freshly formed bond the product has enough time to wait for the saving photon emission.

Reactions proceeding in dense clouds are very much different. Also the type of information differs from the optical absorption spectra of the dilute clouds. There is too much material in a dense cloud for the light to pass through, similar to the sunshine's inability to travel through fog. Fortunately, short wave radio signals of the millimetre-centimetre range can penetrate even such high density material. The energies of molecular vibrations and rotations being much below the energies of electron excitations (cf. Fig. 32 in Chapter "Between Chemistry and Physics"), they coincide with the photon energies of short wave radiation of the frequencies in the $\omega \sim 80$–100 GHz range. By making use of microwave spectroscopy the rotation-vibration landscape of a molecule can be drawn.

Moreover, dense clouds consist mainly of hydrogen in the form of H_2, and helium. Cosmic radiation is absorbed by these substances. The first stage of the majority of chemical reactions is their ionization,

$$H_2 \xrightarrow{\text{radiation}} H_2^+ + e^- \tag{4}$$

The molecular hydrogen ion, H_2^+, enters the reaction,

$$H_2^+ + H_2 \rightarrow H_3^+ + H \tag{5}$$

followed by H_3^+ reacting with atoms of other kind,

$$H_3^+ + O \rightarrow OH^+ + H_2 \tag{6}$$

Both proton and H_2 may undergo further reactions, for example in the presence of CO formaldehyde, H_2CO might also form. Larger molecules may get synthetized similarly.

It is far from certain that the above synthesis routes are the real ones, other ways can also be conceived of. The main thing is that indirect radiation chemical reactions proceed, resulting in the formation of larger molecules in dense clouds. Radiation induced, indirect chemical reactions have been thoroughly investigated in terrestrial laboratories. For example, if dilute aqueous solution is irradiated with fast electrons the total radiation energy is absorbed by water, bringing about ions, atoms or excited water molecules, which react with the dissolved compounds. A good part of our organism can be regarded as aqueous solutions; the biological effect of ionizing radiation can be attributed to indirect radiation chemical actions.

Still, numerous reactions proceeding in the cold of the outer space require further explanation. Activation energy is necessary for the great majority of reactions, Fig. 2 in Chapter "Which Way and How Fast". is a reminder of that. Activation energy is to be overcome by the molecules through thermal motion. If the temperature is low the molecules cannot pass the hurdle. Indeed, the great majority of chemical reactions slows down markedly with decreasing temperature.

Some decades ago, however, certain reactions were observed, whose rates did not decrease with decreasing temperature anymore, after having passed a limit. If the temperature is below 20–40 K reaction rate is independent of temperature. One example out of many is the reaction between methyl radical, $\cdot CH_3$ and ethyl alcohol, C_2H_5OH, the dot symbolizing a "free valence", i.e. unpaired electron,

$$\cdot CH_3 + C_2H_5OH \rightarrow CH_4 + \cdot C_2H_4OH \tag{7}$$

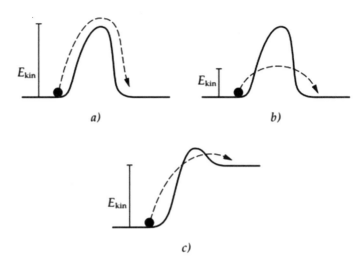

Fig. 4 Tunnel effect **a** a macroscopic ball surmounting a hill, **b** an atomic particle surmounting a potential barrier, **c** energy conservation remains valid for atomic particles

The energy threshold shown in Fig. 3 in Chapter "Which Way and How Fast". does not disappear. But it comes to light that molecules are too small to apply the classical laws of motion without any limitation. A rolling ball can reach the valley beyond the hill if its kinetic energy is higher at the beginning than its potential energy at the hilltop. This is drawn in Fig. 4a. Atomic particles behave in a different way. Both experiments and quantum mechanical inferences show that an atomic particle or an electron can pass a potential threshold although it has less kinetic energy than the height of the barrier. To formulate it more precisely, if the kinetic energy of an atomic particle is less than the height of the potential barrier eventually it might overcome that. If, however, its energy is higher than that, eventually it might not overcome that. No certain statement can be made, only probabilities can be evaluated. Although high kinetic energy results in a high probability of getting through the obstacle, neither advantageous energy relationships make success certain, nor disadvantageous ones doom the attempt to pass a failure, as it is symbolized in Fig. 4b.

Nevertheless, conservation of energy must not be violated. If the final state is of higher energy than the initial one the particle can get through only if its kinetic energy is not lower than the energy difference of the states, as it is given in Fig. 4c. What was discussed above refers only to the transformation of kinetic to potential energy; total energy is impossible to be produced. Only one kind of energy can be transformed into another. This is obviously true also in quantum mechanics.

All that was described above is called tunnel effect since it appears to our eyes trained on macroscopic objects and events as if there were a tunnel for the particle to cross the hill. There is, of course, no such thing at hand neither can any other graphic interpretation be given for the effect. All that can be stated is that evaluating the probabilities of a particle to reside on either side of the hill, one obtains the above results in agreement with experiments. There is nothing more to be said.

Applying these ideas to the present problems, it becomes obvious that chemical reactions might proceed at extremely low temperatures. Reactants can overcome the activation energy barrier due to tunnel effect. A number of laboratory experiments show that this may explain the formation of large molecules at the inclement temperature of the outer space.

Further Reading

C.R. Cowley, *An Introduction to Cosmochemistry* (Cambridge University Press, 1995)

A. Goswami, B. Eswar Reddy (eds.), *Principles and Perspectives in Cosmochemistry*, Lecture Notes, Kodaikanal Observatory, India (2008)

H.Y. McSween, G.R. Huss, *Cosmochemistry* (Cambridge University Press, Cambridge, 2010)

Hydrogen and Energy

1 Some Questions About Near-Future Energetics

Many centuries ago, England was covered with thick forests. However, expanding population demanded more food and the agricultural output could be increased only through extensive farming under medieval conditions. Arable land was to be enlarged, so forests were to be cleared. Industrial revolution only aggravated the problem because more and more iron and steel was needed. For millennia iron ores have been reduced to iron by charcoal and the process has been based on the following reaction,

$$2Fe_2O_3 + 3C \rightarrow Fe + 3CO_2 \tag{1.1}$$

Wood, if heated in a charcoal kiln closed from the open air (under almost anaerobic conditions), gets charred and can be used in iron ore smelting. Iron production contributed in a good part of the English hills becoming barren of trees.

It happened in 1709 that an English artisan, Abraham Dudley realized that coke produced of coal can be used for the reduction of iron ores. From that time coal mines took the role which was played earlier by woodlands. Apart from its great importance in iron production, this was the start of a new age of energetics. The use of wood is tantamount to the use of solar energy in view of the plants' photosynthesis. The age of fossil fuels started with the opening of coal mines and producing coke. Today only some 2% of our consumed energy comes from wood.

It is the history of the near past and of our present time, as oil and natural gas have replaced coal, not in iron smelting but in energetics as a whole, although the relative weights of the different fossil energy carriers are exposed to the ever changing production and prices. Short-term predictions on world economy are beyond the scope of this book. However, there is no doubt that in the long run the world will run short of both gas and coal, faster of the first, slower of the second one. Still, the pressing need to find new energy sources is not a direct consequence of the limits on fossil resources, much more of the state of our global environment, the dangerously contaminated atmosphere, the unbalanced climate, and the polluted oceans.

Two possibilities are offered by nuclear physics. One of them has been real since the middle of the previous century. The energy liberated at the fission of uranium or plutonium nuclei has been made practical of use by the nuclear power stations all over the world. This technique, by now regarded as traditional, has gained an important but not overwhelming share in the global energy mix. (This is the technical term for the world energy consumption as composed of different sources of energy). At present, more than 84% of the global energy need is covered by fossil fuels and only 4.3% comes from nuclear fission. Considering electric energy production alone, a little more than 10% is of nuclear origin. Admitting large national or regional differences fossil fuels still hold their sway over energetics.

The other way to utilize nuclear energy is still in the phase of expensive and long drawn out experiments. It consists in the fusion of light nuclei of two hydrogen isotopes, deuterium, ^2D and tritium, ^3T. Both fission and fusion develop huge amounts of energy at high temperatures and in comparatively small volumes. The rates and densities of energy release are much higher than those wonted with chemical sources, burning coal, oil or gas. This is the reason, together with high investment costs and some complicated equipment, why it is impossible to produce nuclear energy for small-scale users on the spot. Production and use of energy are necessarily apart in space and often also in time with nuclear sources. This is unusual in certain areas. The kinetic energy of an internal combustion engine car is there at the moment when the gas is burned.

There is nothing new in the sites of energy production and consumption being remote. It is contemporary with the electric industry. Power station and consumer are often at distances of several hundred miles, high voltage transmission lines being there to bridge these distances. Whereas electric energy transport has become routine, there is one aspect which stems from the nature of electricity and that is grid loss. A fraction of transported electric energy gets lost to the environment in the form of heat. Let us recall what was written about irreversible processes, friction, heat and entropy! Grid loss is an example of those laws.

If current of intensity, I flows through a conductor of resistance, R an amount of heat, P_{heat} is developed during unit time, given as

$$P_{\text{heat}} = RI^2 \qquad (1.2)$$

This is called Joule heat. The longer a transmission line is and the higher R is, the more energy is lost to the environment. The loss is low if the current is kept low by increasing the voltage of the line. The total electric energy transported during unit time, P_{el} i.e. the electric power is given by the product of current intensity and voltage, V,

$$P_{\text{elec}} = VI \qquad (1.3)$$

Comparison of the two equations above shows clearly that it is expedient to use high voltage for a given electric power because this means a low current and, consequently, small heat loss. Obviously, there are certain technical limits on voltages, and the high voltages must be decreased, according to the demands of the user. The 750 kV of the long distance transmission line must be transformed to 220 V for the sake of our toaster.

Nature offers other possibilities beyond the fossil and nuclear options. Renewable sources not burdening the environment appear to be most attractive to both the policy makers and the general public looking with dread and terror on smoke, carbon dioxide and radioactive waste. Dread and terror being completely justified, renewables certainly seem to offer a good way out of the grip of the contradictory demands, high living standards and low energy consumption. Large steps forward can be seen worldwide in the development and utilization of solar and wind energy equipment. Despite great and successful efforts, the contribution of the renewables is still limited. There is a good deal of work ahead to be done, we will come back to that later.

By renewables such sources are understood where the period of energy accumulation is commeasurable with that of its consumption. Without this limitation, coal or gas would also be classified as renewables, with the only difference that their accumulation takes millions of years and their exhaustion is a matter of a few centuries or even several decades. Obviously, present-time solar energy utilization is not based on photosynthesis and slow carbonization but the task is to make direct use of the energy of oncoming heat and light.

Some relevant methods will be dealt with later. Here we pay attention to the difficulty which is a consequence of the periodicity and of time shift between energy production and consumption. Solar power varies not only with day and season but also with changing meteorological conditions. Consumption cannot allow for all these. Light bulbs are needed at night and trains must run round the clock. The success of the renewables lies with the implementation of industrial scale energy storage.

Energy storage consists in doing work on an appropriate system bringing it into an energy-rich state to the expense of the source. Appropriate means that the system stays in its energetic state for an extended period of time. Now, if the stored energy is needed the system must perform work. One of the practical methods is pumping water into a high lying reservoir doing work against gravitation. Water keeps its potential energy as long as the locks are closed. When energy is needed water is made to flow downwards, operating a turbine which produces electric energy. Electric energy is stored in that way, however, indirectly. First electricity drives the pump, thus, it is transformed into mechanical energy, then the mechanical energy of the turbine is transformed into electricity. But all these steps go with losses due to friction and heat dissipation. The Second Law of Thermodynamics cannot be circumvented.

2   Direct Storage of Electric Energy

Electric energy can be stored without taking recourse to mechanical motion. Some work, however, must be done anyhow since, as it has been clear, heat is the only form of energy that can be transported directly. Let us now consider methods where electric work does the job. The sketchy scheme is seen in Fig. 1. Energy production starts with a turbine (not shown) that drives a generator producing electric voltage. Voltage makes electric charges move, work being done if the distances between opposing charges, which attract each other, are increased. That is the way the mechanical energy of the generator is transformed into electric energy. Separated charges can be collected on two parallel metal plates, positive charges on the one, negative charges on the other plate. If the connection of the pair of plates, called a condenser, is severed from the generator the charges stay separated, in an ideal situation for infinitely long, similarly to water in a hilltop reservoir. Connecting a user, e.g. an electromotor or a light bulb to the plates, the charges move towards

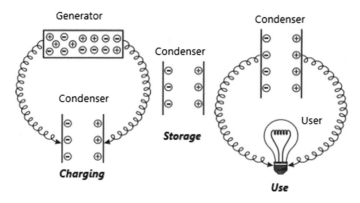

Fig. 1 Electric energy storage by a condenser

each other, current flows across the user. The stored electric energy is being made use of.

The amount of stored energy is given by the amount of charges, Q and voltage, φ,

$$U_{\text{electric}} = \frac{1}{2}Q\varphi \qquad (2.1)$$

Charge and voltage are not independent quantities, the higher the voltage applied to the condenser the more charge is stored,

$$Q = C_e\varphi \qquad (2.2)$$

The proportionality factor, C_e is called capacity, it increases with the area of the plates, A, it decreases with their distance, d and also depends on the insulating substance between the plates, whether it is vacuum, air, wax, mica, or... The effect of the capacitor is characterized by its relative permittivity, ε and the capacity is given as

$$C_e = \frac{\varepsilon A}{d} \qquad (2.3)$$

The relative permittivity of vacuum is 1, of mica is about 8, of water at room temperature 80; a value higher than 100 is a rarity among simple substances.

Much energy can be stored if both Q and φ are large, that is C_e should be large. Large plates with small distance between them and high relative permittivity are desirable. It is difficult to achieve these with pieces of metal. A metal surface in contact with an aqueous solution, however, behaves as a well-contrived condenser. Water and several other liquids containing dissolved salts conduct electricity. Such a solution is called electrolyte.

Consider the two, much different substances, a metal and an electrolyte! Electricity is transported in a metal by the electrons, the positively charged metal ions can just vibrate in their crystal lattice. In the electrolyte, however, both positive and negative ions move freely, transporting electric current. At the metal/electrolyte boundary the mode of current transport changes abruptly. Dissolved ions take or give electrons to the metal, thus ensuring current propagation. In the absence of any current flow the charge carriers brought about by carrier exchange do not leave the boundary layer, their densities becoming different from those in the bulks of the phases: of the electrons in the metal, of the ions in the solution. This situation is very similar to that of a charged condenser with the important difference that both plates of a metal condenser are populated by electrons. In an electrochemical system, shown schematically in Fig. 2, the electric potential, φ is determined by the varying ion densities at the solution side of the interface.

With no current flowing, metal and solution are in equilibrium. Also, here it is valid that free energy, A is at its minimum in chemical equilibria (cf. the chapter on double arrow). But now also an electric energy contribution is to be taken into account. As a result of this a well-defined potential gets established between liquid and metal if they are in equilibrium depending only on concentrations and temperature.

Let a lead sheet be immersed into a sulphuric acid solution of lead sulphate. (Wanting to end up at the lead battery we do not propose a haphazard example.) If the solution is dilute in $PbSO_4$ lead is going to be dissolved. However, no lead atoms exist in water, they must be positively

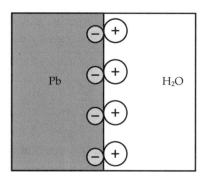

Fig. 2 Schematic view of a double layer at the boundary of metal and aqueous solution

charged forming lead ions. Simply said, lead gets dissolved as lead sulphate,

$$Pb + H_2SO_4 \rightarrow PbSO_4 + 2H^+? \tag{2.4}$$

and the salt undergoes electrolytic dissociation,

$$PbSO_4 \rightleftarrows Pb^{2+} + SO_4^{2-} \tag{2.5}$$

The question mark at the end of the first equation refers to a serious mistake, something has been forgotten. Particles with no electric charges figure on the left hand side, whereas two positive charges appear on the right hand side. But electric charges cannot be created by any process, chemical or otherwise, conservation of electric charges is a basic law of nature. Also, two electrons must be considered that remain in the metal as ions leave the solid phase. Let the equation be corrected as

$$Pb_{metal} \rightleftarrows Pb^{2+}_{solution} + 2e^-_{metal} \tag{2.6}$$

As metal is being dissolved, there is an increase of ion population in the solution and electron population on the metal surface. Due to mutual attraction ions and electrons gather at the interface. There forms an electric double layer with its positive side towards the liquid, negative side towards the metal as it is seen in Fig. 2. The double layer as suggested by the figure and as it exists also in reality is very reminiscent of a charged plate condenser the two plates corresponding to metal electrons and dissolved ions, respectively. The important difference is that whereas any voltage can be switched to the condenser the double layer voltage is determined by the chemical equilibrium.

The distance between the layer being in the order of atomic distances is about 10^{-9} m. It would be impossible to put two isolated metal plates as close to each other as that. Using an electrolytic double layer instead of a metal condenser for electric energy storage the stored energy is much higher simply because d is small.

That is the basic idea behind the lead battery, and all further batteries and dry cells. As to the lead battery, one of the plates is covered with PbO_2 the other consists of pure lead. Lead has four positive charges in the dioxide. The battery is shown in a charged state in Fig. 3. Lead dioxide is positive with respect to the solution because lead goes over to the solution as Pb^{2+} ions, leaving two positive charges behind,

$$Pb^{4+}_{solid} \rightleftarrows Pb^{2+}_{solution} + 2\oplus_{solid} \tag{2.7}$$

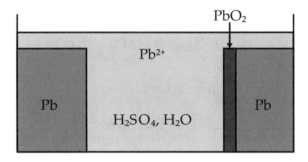

Fig. 3 Charged lead battery

The two equations above describe how the lead battery works. If negative charges are given to the lead plate and positive ones to the lead dioxide both reactions proceed along the lower arrow, metallic lead being deposited on the one, lead dioxide on the other plate. That is battery charging. If the plates of the charged battery are connected by a user (see Fig. 1) both reactions follow the upper arrows Pb^{2+} ions forming at both plates creating a current which flows opposite to the charging one. That is the way stored energy is released. Positive and negative charges that were separated as the battery was charged recombine, hence their energy decrease. To the benefit of the user.

The idea of the method is clear and it works in practice all right in most of the cases (usually the car can be started) but the amount of charged electricity is very limited. Voltage V depends on the composition of the battery and C_e is difficult to influence. Storage time is certainly not infinitely long because reactions other than the ones which are useful in charge storage and current flow inevitably take place. Lead sulphate is produced from lead and lead dioxide along these reactions without the battery producing a current.

It would be most desirable to produce capacities, similar or even larger than those of the electrochemical double layers, without the limits stipulated by chemical equilibria. That is the aim of the supercapacitors developed in the recent decades. Over centuries it was a part of common knowledge that elementary carbon has two allotropic forms, diamond and graphite. (Different crystalline structures of an element or compound are called allotropic forms). Investigations of the last decades revealed the existence of further crystalline carbon structures. At first, the nearly spherical molecule fullerene of the composition C_{60} and its derivatives were found. Later carbon nanotubes were obtained, finally graphene layers, which correspond to a single layer of the hexagonal graphite have become available (Fig. 4). Graphene, known for quite some time, was most cumbersome to produce,

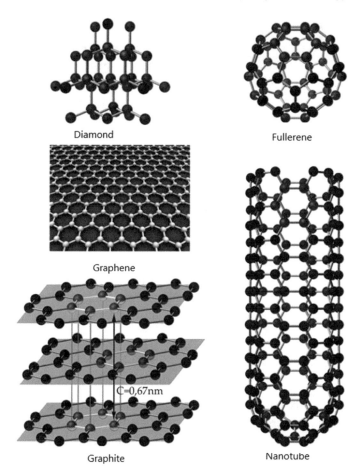

Diamond

Fullerene

Graphene

Graphite

Nanotube

Fig. 4 Carbon modifications (based on Wikipedia)

hence it was very expensive and little was known about its favourable properties. Two researchers, *Geim* and *Novoselov*, working in Manchester proposed
a strikingly simple method, sticking a sellotape to a graphite crystal and skin-
ning its surface by pulling tape and crystal apart. This method opened up the
research activity on graphene (and resulted in a Noble prize to the two men).
Graphene has become an integral part of transistors, electrode materials and
solar cells by now.

Here we are going to deal with one single application of the modifications of carbon. By making use of nanotubes or graphene, structures were
developed whose capacities are orders of magnitude beyond the traditional
metal/insulator systems and also of electrochemical double layers. Called

supercapacitors they consist of a carbon nanotube or graphene layer sandwiched between two metal electrodes. The thickness of the layer, d is about several times 10^{-9} m and its surface A is large enough to cover three or four tennis courts if one gram of carbon were smoothed out in a plane. This immense area is due to the porosity and multiple corrugation of the substance.

The electrical double layer in the graphene soaked with an electrolyte behaves as if it were made of an insulator of extremely high relative permittivity, at least in a certain voltage interval. There are huge capacities, C_e values at hand whereas the voltage V is independent of the construction of the storage device in contrast to the batteries.

The device seems to be ideal, but it still has a serious disadvantage, it can maintain the charges only for a limited period of time. As far as long-time storage goes, batteries still have their dominance. If, however, a large amount of energy must be stored for relatively short periods and then used for fast battery charging, as is the case with electric vehicles, supercapacitors have turned out to be more useful.

3   Hydrogen Energetics

The energy industry of the world is a fast changing field. Our endangered environment demands carbonaceous fossil sources: coal, oil and natural gas to be phased out. Some decades ago we were threatened that their sources will run out soon. Now it is the changing climate with all its consequences that prompts us to reduce the use of these materials.

At present, the world is only at the very start of this transformation. Still more than 84% of global primary consumption comes from fossil fuels, 11.4% from renewable sources (hydropower, wind, solar, biofuels) and 4.3% from nuclear fission carbonaceous materials still being in the lead. The energy mix must be changed radically! It is still not known whether the solution lies with nuclear fission, renewable sources or a clever combination of the two with nuclear fusion being still a remote hope. It seems, however, to be sure that whatever sources will be chosen new ways and methods for the transport and storage of energy must be found. It is generally held that to this end hydrogen is the material of the future. As a carrier but not as a source!

It is the first time in the history of engineering that source and carrier of energy become different. Firewood, coal or natural gas are sources which store energy and serve also as carriers (when coal is carried to the cellar energy is transported). Hydrogen is different because no hydrogen fountains exist in nature. It would be, however, most desirable to transform the very high energy density of the hazardous nuclear sources or the very low energy density of the whimsical renewable sources into the modest chemical energy of the safe hydrogen. Hydrogen stores energy for an infinite period of time, offers easy ways of transportation and can be applied in a number of different ways.

The advantages are obvious. Hydrogen can be produced from water, the most abundant substance on the surface of the Earth. Energy storage is effected by water decomposition,

$$H_2O \rightarrow H_2 + 1/2\,O_2 \qquad (3.1)$$

a process which requires energy, part of which becomes the chemical energy of hydrogen. Chemists have been well versed in these methods, however, which are both economic and environmentally friendly, they can also be performed in large volumes and are not based on fossil materials. They are, however, still to be developed further.

When hydrogen is being burnt its energy is totally released in the form of heat. (Totally, yes, Second Law sets a limit only for producing work from heat; heat from work can be obtained with no limitations.) This process certainly does not harm the environment, its only product being pure water.

Hydrogen can store energy for an unlimited period of time without any loss because it is of negligible reactivity at ordinary temperatures and can do so in the absence of any catalyst. There is no process by which energy would leak. It is of course explosive if, mixed with oxygen, a spark jumps over or is contacted by naked flame. In that case its use is as risky as that of petrol or natural gas. Nuclear fission materials and wastes should not even be mentioned in that context.

Nevertheless, one must not forget about two serious hydrogen-related accidents. The first of the two destroyed a hydrogen filled Zeppelin-type airship. Airships were expected to play an important role in air traffic at the first part of the twentieth century. These vehicles were kept afloat by low density gas, hydrogen or helium, and were propelled and steered by motor driven propellers. They plied more or less regularly between Germany and the Americas. At New Jersey, on May 26, 1937 hydrogen filled airship Hindenburg went up in flames at the moment of landing and burnt to ashes in minutes. It happened in modern times so radio stations could report from the spot and newsreel cameras recorded the event. The primary cause of the accident is still debated. It might have been a spark by friction electricity that ignited the vessel and perhaps it was not so much the hydrogen but the textiles that were burning. All these debates were, however, in vain, hydrogen-filled Zeppelins were doomed.

The second accident happened closer to our time. Airship Challenger launched in the US on May 28, 1986 and exploded several seconds after its start. The reason was an O ring which, having lost its elasticity, let liquid hydrogen escape. Since this took place in the technical age all the details were recorded by TV cameras and were released in real time. Had someone missed the initial broadcast, they could enjoy the repetitions over hours and days.

Hydrogen can be used in internal combustion engines and such types of engines are well-known. As far as safety is concerned, car factories have a particular predilection to a pair of photographs, one of them showing a hydrogen-driven, the other a petrol-driven car; both of them catch fire. The hydrogen flame is seen to be an upward torch which, when extinguished, left the car virtually undamaged, whereas the petrol flame destroyed the car altogether.

Being a very light substance, it is most suitable to fuel aeroplanes. Also, it is much used in spacecrafts; for example, Saturn rockets were fuelled by liquid hydrogen. Obviously, it can be used directly for heating. Its great importance in ammonia manufacturing was discussed in an earlier chapter, at present the overwhelming part of hydrogen is used by that industry. Metallurgy would also be reshaped if cheap hydrogen were available in large quantities. As mentioned above, metal ores are usually reduced with coke a material which is easy to be replaced by hydrogen.

Hydrogen being burnt can produce electricity through heat exchangers, turbines and generators exactly as fossil fuels do. However, the losses due to multiple energy conversions—chemical energy to heat, heat to mechanical energy, mechanical energy to electricity—can be decreased if chemical energy is directly converted into electricity. These methods will be discussed later.

Also, the sun's radiation can be exploited to produce hydrogen. This is a very important aspect since energy storage is a condition of the use of solar energy—energy gained during the shiny hours must be stored for the dark periods. One of the most important properties of a storage material is the amount of energy stored in its unit mass. Batteries made of heavy lead and dense sulphuric acid fare badly in this comparison. Given its high heat of combustion and low density hydrogen is the most appropriate substance in this respect. One kilogram hydrogen produces three times more heat than one kilogram liquid hydrocarbon.

Storage and transport of hydrogen are mostly successful, although there are still some problems to be solved. Gaseous hydrogen, despite its boiling point being as low as 20 K ($-253\ °C$) is liquefied and the liquid stored at industrial scale. Tanks of several thousand cubic metres have been built where the evaporation loss is low thanks to good isolation and small free surfaces. Liquid hydrogen can be transported by rail or by ship. Although it takes much room because its density is a mere 7% of that of water this is counterbalanced by the high heat of combustion: a tanker filled with liquid hydrogen carries more energy than another one filled with oil.

Pipelines can be used also for hydrogen transport, obviously as a gas. Low density is here a disadvantage because the energy driven through a given cross section is low, a fact which is partly counteracted by the low viscosity of hydrogen, due to which compressors have an easier job making hydrogen gas flow.

Small amounts of hydrogen can be closed into steel cylinders, although this method of transportation is not too expedient in view of the cylinder weight. There is another, most promising method for small or medium amounts of gas storage and transport. Some metals and alloys absorb large amounts of hydrogen if the gas pressure is high and release it again if the pressure is low. This is an attractive method with some pitfalls which will be discussed.

Given all these advantages the question appears naturally, what is the reason why this excellent fuel has not come into general use yet. On the one hand, there is a lot of fossil fuel at our disposal and environmental concerns still weigh little. On the other hand, energy industry with all its expensive investments, reacts most carefully. Besides, the hydrogen industry is still hampered by a good number of chemical and engineering problems unsolved. This will be discussed below.

Nevertheless, hydrogen energetics is emerging. As an example, the European Union made plans to produce 1 million tonnes of hydrogen by 2024 and 10 million tonnes by 2030 through electrolysis. This seems to be a great step forward.

4   Hydrogen Production in the Future Continuous Tense

As it has repeatedly been mentioned in this book hydrogen gas is no newcomer in the chemical industry. Ammonia manufacturing and the oil industry require a huge amount of hydrogen, which is produced from water by making use of fossil sources, coal or nowadays overwhelmingly methane (see the chapter *How to produce ammonia*). The fossil material acts in these processes both as a reactant and as an energy source. The chemical industry, in an endeavour to get rid of fossil materials, tries hard to apply alternative energy sources and find new reactants for hydrogen production, a substance to be used much beyond the needs of ammonia production and crude oil treatment.

4.1   Water Decomposition by Heat

Most regrettably, hydrogen wells do not exist in nature. Hydrogen must be made from water, a source which is unfortunately not exhaustible. Some time ago the amount of water was estimated in comparison with the need if all vehicles of the US were driven by hydrogen. Given the water resources of America, this would not pose any problem. However, the water economy of developing countries with limited fresh water supply might be shaken. A good part of these countries are rich in sunshine, which is the hoped-for primary energy source of the future. Hence energy and raw material could be exchanged between different parts of Earth. Water for sunshine—a curious world!

Water decomposition into its elements requires energy. Being an endothermic process, it is enhanced by high temperature (recall the Le Chatelier-Brown principle!). Future nuclear reactors will work at much higher temperatures than the present ones. Nowadays, reactors are cooled by water, which is used also for heat transport. In the future, sodium or lead melt, helium gas, perhaps supercritical water will be used. (Supercritical water is no new substance, it is ordinary water at a temperature so high as to make vapour impossible to be condensed at whatever pressure; supercritical vapour behaves as a permanent gas. This state is shown by the top curve in Fig. 15 in Chapter "Between Chemistry and Physics".) Such substances will carry the heat developed in the active zone of the future to the turbines where thermal energy gets transformed into kinetic energy. Consequently, the carrier liquid cools down. The larger temperature difference prevails between source and sink of heat the higher percentage of heat is transformed into work. This is a consequence of the Second Law of thermodynamics or to be more faithful to history, the Second Law was based on the study of the efficiency of the steam engines revealing the fact that work cannot be made from heat without limit.

The warmest point of a nuclear reactor is its active zone of temperature T_z; this site is the source of energy, U_z which is proportional with T_z, $U_z \propto T_z$, since heat is the sole form of energy developed in the zone. Temperature difference is needed for heat to be transported. If the lowest temperature in the turbine is T_t the heat given to the turbine from the zone is proportional with the temperature difference. This amount of heat can be transformed into work, L, that is $L \propto (T_z - T_t)$. The efficiency, η is the ratio of work and of the energy coming from the source,

$$\eta = \frac{L}{U_z} = \frac{T_z - T_t}{T_z} \qquad (4.1)$$

The larger the difference between the temperatures of zone and turbine the higher the efficiency: the larger fraction of the heat developed in the zone can be turned into work. Thus, it is expedient to release energy at a high temperature, an aim which has its technical limits. Obviously, what was written here is not limited to any mode of energy production. The above equation describes the efficiency of any heat engine. It is called Carnot efficiency, taking its name from the first person who realized it, *Sadi Carnot*, a French military officer (Fig. 5).

One of the most important questions regarding modern hydrogen production is whether the heat of non-fossil sources can be harnessed directly for water decomposition. The use of higher and higher temperatures is demanded in view of Carnot efficiency, whereas it is made possible by the recent developments of material science. A number of processes in industrial chemistry require high temperatures. Thus, this gives some hope towards the practical implementation of direct thermal water decomposition. First, water must be decomposed,

$$H_2O \xrightarrow{\approx 3000K} H + OH \qquad (4.2)$$

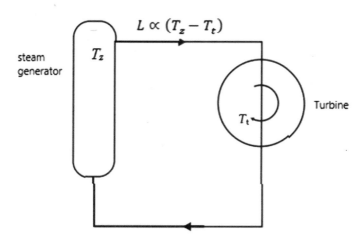

Fig. 5 Large thermal differences increase the efficiency of the heat engine

and after extremely fast cooling the products can be obtained,

$$2H \rightarrow H_2 \tag{4.3}$$

$$2OH \rightarrow \frac{1}{2}O_2 + H_2O \tag{4.4}$$

The first step takes place at very high temperature which might pose a problem even today. But apart from that the direct method is not favourable energetically either. Both H atoms and OH radicals unite with energy release, thus heat is made from heat; despite high temperature only a small fraction of heat can be made use of. A further problem would be the separation of hydrogen and oxygen after the reaction is completed.

Energy conditions are given schematically in Fig. 6. Both H atom and OH radical are richer in energy than the molecules H_2 and O_2. It is of no use ascending a high lying terrace if it is inevitable to climb down again.

The best thing to do would be to transform water in one single step into H_2 and O_2, according to the path of the thick arrow in Fig. 6. Such a process, however, does not exist. Instead, researchers have been trying to find a reaction path that results in molecular hydrogen, while intermediate steps are possibly not much richer in energy than the final product. The dashed line indicates the imaginary energy path of such a desired process. The scheme resembles the energy scheme in Fig. 3 in Chapter "Which Way and How Fast", where the energy barrier was called activation energy. Now we are talking about something similar. The difference between Figs. 3 in Chapter "Which Way and How Fast" and 6 are twofold: the activated state is well known, in the present case they are H and OH; and the final state is of higher energy than the initial one. As to this latter difference, this is the aim of the exercise. That is how energy is accumulated as chemical energy.

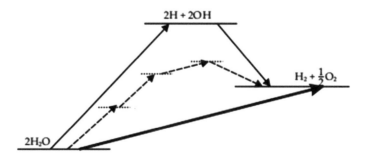

Fig. 6 Energy relations of water decomposition

It is quite usual to try to find a bypath along which the activation energy is lower than that of the direct transformation. The method, called catalysis, does not result in a higher amount of product, but the process becomes faster; this was already mentioned in context of ammonia production. The dashed line in Fig. 6 represents the energy steps of a catalytic reaction for water decomposition. The main point here, however, lies not with the reaction rate but with the energy balance—we are going to economize with the energies of the consecutive steps, whereas the net energy change of the overall process is fixed.

Thermodynamics is a good guide for finding the appropriate reactions. The notion of free energy, A was introduced as ammonia production was discussed. It is the difference of two quantities, the total energy change in the course of the process and the heat exchanged with the environment. As it was written,

$$A_{initial} - A_{final} = (U_{initial} - U_{final}) - T(S_{initial} - S_{final}) \qquad (4.5)$$

Deducing the heat expressed in terms of entropy and temperature from the total energy change one obtains the energy which avoids dissipation. Thus A is nothing else but that part of the total energy change which is the work done by the system (if its energy decreases), or the work done for the system (if its energy increases).

Hydrogen production from water implies energy increase, A is positive. Economy demands finding a path along which A is possibly low. Dissipated energy plays an important role here. If heat is absorbed along a reaction step, called endothermic, the entropy increases, $S_{initial} < S_{final}$ prevails. In that case, heat is added to the energy of the system so it is advantageous if it is large. Consequently it is good if this step proceeds at a high temperature to make the heat $T(S_{initial} - S_{final})$ high. Another step, called exothermic, releases heat, $S_{initial} > S_{final}$ in that case. Here low temperatures are preferable to make the absolute value of the negative product as small as possible in order to reduce energy loss in the form of heat.

In order to make water decomposition energetically favourable one has to find a reaction path where the endothermic steps take place at possibly high, the exothermic ones at low temperatures. The resultant chemical change should solely consist in the formation of hydrogen and oxygen from water, retrieving all other reagents in their original forms and quantities. In one word, water should be catalytically decomposed. The series of reactions should be like the following,

$$2AB + 2H_2O \rightarrow 2AH + 2BOH \tag{4.6}$$

$$2BOH \rightarrow 2B + 1/2\,O_2 + H_2O \tag{4.7}$$

$$2\,AH \rightarrow 2A + H_2 \tag{4.8}$$

$$2A + 2B \rightarrow 2AB \tag{4.9}$$

where AB is the catalyst.

The resultant chemical change can be obtained by treating the reaction equations similarly to algebraic equations: right hand side and left hand side formulae must be added separately and the compounds that appear on both sides must be cancelled. Since a substance figuring both on the left hand side (reactant) and on the right hand side (product) indicates that this substance, although it participated in the reactions, finally gets reconstituted. Performing that elementary algebra with the reactions above one finds the net chemical change to be

$$H_2O \rightarrow H_2 + 1/2\,O_2 \tag{4.10}$$

It happened quite some time ago that all the relevant compounds and thermodynamic data were numerically analyzed, finding about ten thousand reaction paths which more or less complied with the above requirements. Then investigating practical viability, it turned out that the most promising routes are the same as the ones predicted by the chemists' experiences and intuition—to the great satisfaction of conservative minded researchers.

Thermodynamics renders information only about possibilities excluding processes which certainly do not proceed but cannot predict the rates of the reactions. Slow processes are of little industrial use, thus kinetics must also be considered. The slower a reaction is the larger reactors are needed for the production of a given amount. Also, the prices and availability of the reagents, the complexity of the equipment, the environmental damage must be considered. All these must be translated into cost and benefit since what is being produced must be sold. Economy makes the final decision.

The first method which appeared to be realistic from the industrial point of view was based on $CaBr_2$ and mercury. Water is decomposed here by $CaBr_2$

as

$$CaBr_2 + H_2O \xrightarrow{730\,°C} Ca(OH)_2 + 2HBr \qquad (4.11)$$

Hydrogen, the product wanted, develops when HBr is contacted with mercury,

$$Hg + 2HBr \xrightarrow{250\,°C} HgBr_2 + H_2 \qquad (4.12)$$

Finally, in a reaction of two products one of the reactants is recovered,

$$HgBr_2 + Ca(OH)_2 \xrightarrow{200\,°C} CaBr_2 + HgO + H_2O, \qquad (4.13)$$

whereas the other one is retrieved though thermal decomposition,

$$HgO \xrightarrow{600\,°C} Hg + \frac{1}{2} O_2 \qquad (4.14)$$

Actually, the last of the reactions is the one by which oxygen was discovered. The net change of this reaction series can be found as it was shown in the general case. Treating the reactions as algebraic equations one finds nothing else but water decomposition. The above process is somewhat more complicated than the general scheme given above since here figure two compounds as catalysts.

The process has a number of advantages and disadvantages. It is good that the maximum temperature needed is not much above 700 °C and gas cooled fission reactors will produce such a temperature easily, once they come into operation. The products are easily separated due to their different physical properties. Above all, it is free from any side products. Its main disadvantage is the use of mercury, which is both expensive and most dangerous to humans and the environment. A further problem is the large amount of mercury the circulation of which increases operational costs.

Whereas these are problems of ecology and rheology there are further questions regarding chemical kinetics. For example, if the $Hg \rightarrow HgBr_2$ reaction can be made fast this goes with a relatively small amount of necessary mercury, decreasing both cost and danger. As a matter of fact, the rate of this process was markedly increased by optimizing the conditions.

Nevertheless, environmental regulations phased out the large-scale use of mercury. This prompted the development of a version of the above scheme where Hg is replaced by a natural iron oxide, Fe_3O_4, called magnetite. Instead of the hydrogen-producing key reaction between Hg and HBr a two-step process is introduced, the scheme of which writes as

$$\text{magnetite} + \text{HBr} \rightarrow \text{FeBr}_2 + \text{Br}_2 + \text{water} \left(\text{at } 220\,^{\circ}\text{C}\right) \qquad (4.15)$$

$$\text{FeBr}_2 + \text{water} \rightarrow \text{magnetite} + \text{HBr} + \text{hydrogen} \left(\text{at } 560\,^{\circ}\text{C}\right) \qquad (4.16)$$

Auxiliary substances magnetite and HBr are seen to be fully recovered, reaction temperatures are moderate and even a resourceful flow system was devised whereby the process can be run continuously with no need to wait until chemical equilibrium is established. Despite all these, industry has not applied the method, mainly because unwanted side reactions decrease the output.

Thermochemical reactions might be combined with electrochemical steps; these so-called hybrid processes often proceed under milder conditions— lower temperatures and pressures—than their thermal counterparts. One of the many possibilities is centred on the univalent copper chloride, CuCl. It is produced from common $CuCl_2$ in a thermochemical process,

$$2CuCl_2 + 2HCl \rightarrow 2CuCl + Cl_2 \left(\text{at } 500\,^{\circ}\text{C}\right) \qquad (4.17)$$

whereas the hydrogen-producing key reaction takes place in an aqueous solution with the assistance of electricity,

$$2CuCl + 2HCl \rightarrow H_2 + 2CuCl_2 \left(\text{at } 200\,^{\circ}\text{C}, 0.5\text{V}\right) \qquad (4.18)$$

This hybrid process and a number of similar ones have been investigated in much detail, some of them even at a pilot plant level.

Technical literature contains long and detailed lists of industrial methods for thermal hydrogen production. This lavish choice is somewhat of a deterrent, one may feel that it would be far better to have only one process, the chemistry of which is well established, can be performed at an industrial scale and is commercially profitable. It is perhaps no wonder that large-scale plans opt for water electrolysis instead of thermal processes.

4.2   Water Electrolysis

Electrolysis has been a traditional and widespread method of hydrogen production. It is based on the spontaneous, partial decomposition of water into charged components, called ions, as it was described in connection with the structure of water,

$$2H_2O \rightleftharpoons 2H^+ + 2OH^- \tag{4.19}$$

Immersing two metal electrodes, e.g. platinum sheets into water and connecting electric voltage to the sheets ions get neutralized and combine into oxygen and hydrogen,

$$2H^+ + 2e^- \rightleftharpoons 2H \rightleftharpoons H_2 \tag{4.20}$$

$$2OH^- - 2e^- \rightleftharpoons 2OH \rightleftharpoons \frac{1}{2}O_2 + H_2O \tag{4.21}$$

Everybody has seen water bubble around the electrodes of a dry battery. Bubble formation, or gas development, shows that the battery voltage is sufficiently high to overcome a voltage barrier that forms right at the outset of the process. As the above two reactions set in the negative electrode gets surrounded by H_2, the positive one by O_2 (Fig. 7). Electrolysis is a reversible process in the sense of the double arrows in the above equations: equilibria

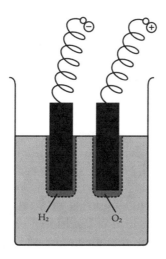

Fig. 7 Gas sheaths form around the electrodes during water electrolysis

are shifted "to the left hand side", i.e. towards the dissolution of the gases in ionic forms as their amounts increase. In that case, electrodes get charges from the molecules the system working as a galvanic cell similar to the charged lead battery. This is a general situation: electrolysis and current generation are opposing processes, both might take place in the same arrangement with the relative concentrations and external voltage determining the directions.

An increase of the electrolysis voltage counteracts the back process: the voltage must be opposite in sign to and larger in its absolute value than that of the H_2/O_2 galvanic cell, defined by the densities of the gases. In that case, the equilibrium is shifted "to the right hand side". The maximum gas density is limited by the atmospheric pressure, were the gas pressure higher than that the gas would escape the vessel. Continuous water electrolysis can be maintained if the absolute value of the electrolyzing voltage is larger than the voltage of the H_2/O_2 cell, both gases being at atmospheric pressure. This voltage is called decomposition potential.

Water electrolysis performed at the decomposition potential seems to be most favourable from a thermodynamic point of view. Water decomposition being an endothermic process the system takes up heat from the environment during electrolysis, thus environmental energy adds to the overall balance. Again, it is the time scale which refutes optimism based on equilibrium thermodynamics. Keeping the voltage near the decomposition potential, electrolysis is much too slow for any practical purpose. Instead, voltage must be increased; although this goes with heat release hence spoils the energy balance, it is favourable to the rate of gas evolution.

Higher temperature is advantageous from more than one point of view. On the one hand, ion neutralization at the electrode surfaces becomes faster; on the other hand, hydrogen and hydroxyl ion concentrations are higher. Water dissociation is endothermic, hence an increase in temperature favours the process. More ions mean lower electric resistance of the solution decreasing the thermal loss as it was shown in context with the power lines.

That is the reason why instead of pure water concentrated aqueous solutions are electrolyzed. Most often solutions containing 20–40% potassium hydroxide are used. Fast charge exchange at the metal surfaces is of prime importance, together with the chemical stability of the electrodes that are usually made of fine nickel powder, zinc or carbon. A porous diaphragm, permeable for liquid hence for electricity, is placed between anode and cathode in order to prevent mixing of the evolved gases. Operational temperature is kept at about 80–100 °C. Traditional electrolyzers produce several hundred m^3 hydrogen per hour at a pressure of 1 atmosphere.

The use of concentrated potassium hydroxide is a permanent source of trouble: its concentration must be maintained and it is rather corrosive. Those are the reasons why contemporary electrolyzers apply conducting polymers, called proton exchange membranes, or PEM, as electrolytes instead of alkaline solutions. The good conductor polymer takes up almost the complete volume of the cell, leaving only thin layers at the two sides for distilled water to flow through (Fig. 8). While it does not attack the cell wall the layers are thin enough not to increase electrical resistance too much. The usual temperature being between 20 and 150 °C an output of about 100 m^3 per hour can be achieved.

Water electrolysis being an endothermic process it is favourable to supply part of the necessary energy in the form of heat. According to calculations, the energy efficiency of water electrolysis is 41% at 100 °C while it is as high as 64% at 850 °C. Economical as it might be to perform electrolysis at a very high temperature, this is no trivial task, ions being non-existent in water vapour. Hence, instead of water, conducting oxides are used as conductors in high-temperature electrolyzers. This arrangement is similar to that of PEM cells with the difference that whereas electricity is transported in PEM membranes by H$^+$ ions, in oxide it is done by O^{2-} ions (Fig. 9).

A very thin layer of water vapour surrounds the oxide, which takes up the greatest part of the cell interior. The poor conductor vapour layer is thin enough to play but little role in the conductance of the cell as a whole. Water molecules dissociate into H$^+$ and OH$^-$ ions at the oxide surface. The oxygen ions of the oxide travelling from cathode to anode take up the charges of

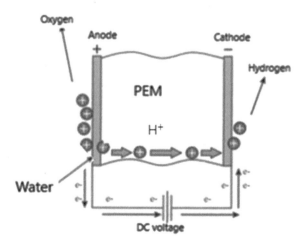

Fig. 8 Polymer exchange membrane (PEM) cell

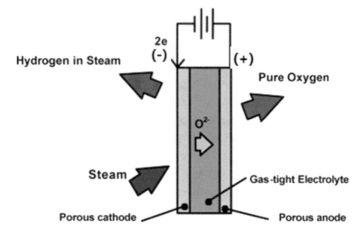

Fig. 9 Steam electrolysis in a solid oxide electrolyzer (after Wikipedia with correction)

the ions which form the gaseous products in the same way as in the case of conventional water electrolysis.

As stated above, water dissociation can be reversed, hydrogen and oxygen can produce current along with water formation. The two gases act as battery electrodes. Expressing the process in terms of charges O_2 and H_2 molecules take up charges forming H^+ and OH^- ions which finally unite, viz.

$$H_2 \rightarrow 2H^+ + 2e^- \tag{4.22}$$

$$\frac{1}{2}O_2 + H_2O + 2e^- \rightarrow 2OH^- \tag{4.23}$$

$$2H^+ + 2OH^- \rightarrow 2H_2O \tag{4.24}$$

A current can be generated if hydrogen oxidation (electron loss) and oxygen reduction (electron gain) are spatially separated. Electron must go from the hydrogen side to the oxygen side along the external circuit through the electric appliance. Such a construction, which is called a fuel cell, is very similar to the one in Fig. 9. The difference lies with the mode of operation. A fuel cell is fed with hydrogen and oxygen (air) and the voltage produced is switched to a consumer: this is the contrary of water electrolysis.

The principle is clear and a number of practical realizations are well known. Just to mention one of them, some electric cars are driven by hydrogen fuel cells. Elementary hydrogen is not the sole appropriate substance, hydrogen-containing organic materials might well serve the same purpose. Some scientists e.g. Nobelist George Olah suggest the use of methanol both as fuel and energy carrier instead of hydrogen. One of its advantages is that methanol can be synthetized from hydrogen and carbon dioxide, a process beneficial for the environment's CO_2 content. In effect this route is still a part of hydrogen economy by storing hydrogen in the form of methanol. Indeed, it is much easier to store and transport methanol than hydrogen. This asset is perhaps greater than the inevitable energy loss required by methanol synthesis.

The principle of fuel cells is clear their technology is being developed fast. Good membranes and efficient electrode materials must be found, which act also as efficient catalysts of the electrode processes.

Several years ago, hydrogen manufacturing by water electrolysis, a feasible technology otherwise, was thought to be poor from the point of economy. The multiple conversions of the energy of a primary source into heat, heat into mechanical energy, this into electric energy which is finally stored as the chemical energy of hydrogen involve serious losses and is worth doing only where electricity is very cheap. The present-day stance, however, has turned much in favour of electrolysis, with a view to carbon-free hydrogen production.

5   Sunlight and Hydrogen

A primary energy source by which water can be decomposed directly—that would be a great boon. It would make hydrogen production both simple and economic. A number of physical and chemical processes effected by light being well known, the radiation of Sun as a primary source is worth considering. It seems to be a sober idea to produce hydrogen by sunlight.

Sun does not glare all over the globe with the same strength, solar radiation is still the most evenly distributed energy source in the world. The yearly average insolation (incident solar energy per unit area) does not vary too much over the globe: the shiniest parts of Africa obtain only three times as much sunlight as the areas near the poles, considering the whole year. The differences lie with intensities, i.e. the sunlit hours per day. Sunlight is no economic energy source under cloudy skies. Still, everybody gets enough of the rays of Sun, or can change sunlight for water as it was mentioned earlier.

This would be the most environmentally friendly energy source. Not only substances can contaminate, one had to learn the notion of thermal pollution, the deleterious temperature increase of water and the environment as a whole. Burning hydrogen, apart from trace amounts of nitrous oxides produced from the air, transforms into pure water. If hydrogen is made at the expense of solar energy the environment stays free of thermal pollution.

If any traditional or modern energy source is used, be it of fossil or nuclear origin, the entire surface of the globe is heated, mostly the area where the energy is liberated or made use of. Both the cooling water of the power stations and the turning tool of a lathe get warm, in accordance with the Second Law as stated by Clausius: *The entropy of the universe tends to a maximum.* The entropy of the heavily industrialized areas and of the Earth surface as a whole certainly increases. This is true even apart from the additional harm due to greenhouse gases.

Sunlight is the only energy source that does not cause thermal pollution. Transforming solar energy into work the amount of heat which reaches the surface is diminished. Finally, this work too is transformed into heat but only that amount of heat is produced which would have reached the Earth anyway.

5.1   Water Decomposition with Solar Heat

Whereas the amount of energy coming from the Sun is high, its surface density, taking light and heat together, is low, about 1 kW/m^2 in the temperate zone. This is not too much, a household immersion boiling device with 300 W power and a surface of some 0.003 m^2 has a hundred times higher surface power density. Hence, sunshine cannot produce temperatures for any practical industrial use. Let us recall Carnot efficiency! This source is adequate only to help out detached houses with warm water in the Mediterranean.

However, solar power density can be increased locally. The idea goes back to Archimedes, who was also a resourceful military engineer and allegedly set fire to the Roman fleet at Syracuse that way. Sunlight collected by large-surface concave mirrors is reflected onto a small area. Energy density can be increased hundred or even ten thousand fold by such a device. Temperatures beyond a thousand centigrade can be achieved near the focal point of the mirror. One out of many practical realizations of such concentrating collectors is given in Fig. 10.

Thermal decomposition of water can be performed at temperatures as high as 2000 °C or above a value which can be attained in a concentrating collector. The price of the complicated machinery and the structural materials which go with such high temperatures poses an obvious economical problem. A seemingly simple solution is based on the thermal decomposition of zinc

Fig. 10 Stirling type concentrating collector. Image by: Schlaich Bergermann und Partner and released into public domain at http://wire0.ises.org/wire/independents/imagelibrary.nsf—This site states explicitly that all images are public domain

oxide,

$$ZnO \rightarrow Zn + 1/2 O_2 \text{ at } 1900\,°C \tag{5.1}$$

followed by the high temperature reaction between zinc and water,

$$Zn + H_2O \rightarrow ZnO + H_2 \text{ at } 427\,°C \tag{5.2}$$

The catalytic cycle is seen to be complete the only products of the reaction sequence being hydrogen and oxygen. Later the thermal dissociation of cerium tetroxide into cerium trioxide, followed by the reaction with water turned out to be more advantageous,

$$2CeO_2 \rightarrow Ce_2O_3 + O_2 \text{ at } 2000\,°C \tag{5.3}$$

$$Ce_2O_3 + H_2O \rightarrow 2CeO_2 + H_2 \text{ at } 400\,°C \tag{5.4}$$

The first, high temperature step is endothermic, the second one, which takes place at a lower temperature, is exothermic fulfilling what is advised on the basis of thermodynamics.

Apart from direct hydrogen production solar heat can generate electricity similar to the heat of fossil or nuclear fuels. The methods of electrolytic hydrogen production were described in the previous section. The differences between the sources of energy pertain economic and environmental problems in that context.

5.2   Water Decomposition with Sunshine

Light is very different from heat in the context of hydrogen production as far as both energy efficiency and chemical reactions are concerned. Light energy is administered to a molecule in high density packages. As it was discussed earlier, a photoelectric effect can be interpreted in terms of the idea that light energy propagates in packages of $\hbar\omega$, called photons, this being the smallest amount of energy which can be emitted by a source and absorbed by a medium. The analysis of the line spectra and the rules of quantum mechanics show that this minimum energy is adsorbed by a single molecule by getting one of its electrons to a higher energy state or, with low energy packages, by making its valence oscillations more energetic.

Photon energy is very high as compared to that of thermal motion at moderate temperatures. It was shown in connection with specific heat that the average energy of a diatomic molecule, e.g. H_2, is $\varepsilon = \frac{5}{2}k_B T$ it being about 10^{-20} J/molecule. In contrast to that, the photon energy of the visible light, e.g. that of the yellow line of the sodium spectrum, denoted as line D, is $\varepsilon_{NaD} = \hbar\omega_{NaD} \approx 3.5 \times 10^{-19}$ J/molecule. A single photon of the visible light delivers thirty to forty times more energy to a molecule than the molecule's share from thermal motion. (The figures used are T = 300 K, $k_B = 1.38 \times 10^{-23}$ J/K·molecule, $\hbar = 1.05 \times 10^{-34}$ J/s, $\omega_{NaD} \approx 3.3 \times 10^{15}$ s^{-1}.)

As far as energy is concerned, does light act as if the illuminated substance's temperature were raised to about 15.000 °C? Well … not quite. Equipartition law, mentioned when specific heat was discussed, states that thermal energy is distributed equally among the rotations and vibrations of the molecule. That means that when the temperature rises, each possible motion of the molecule becomes faster, more energetic. If, however, a photon is absorbed, the energy of one single electron or of a certain mode of vibration is changed, leaving all other possible motions unaffected. Photon energy is not distributed right after absorption but remains, for a while, where it was absorbed. Thus, it might happen that light administers more energy to a molecule than it would be possible to do thermally. The molecule might undergo thermal decomposition if the temperature is much too high. Photon energy, absorbed by an electron, does not reach the weak bond instantaneously. Hence, highly energetic molecules, prone to fast reactions, can be produced by light absorption.

The local accumulation of energy is most important for the practical use of photochemical reactions. Now the question is whether hydrogen can be produced from water by some (or any) photochemical process. The dream would be a direct photochemical reaction where water would be decomposed into its elements without making use of any auxiliary substance. This is obviously impossible, according to everyday experience, nobody has ever seen a hydrogen bubble rise from a sunlit lake. A glimpse at a glass of water tells us why: water is colourless, light oncoming on one side of a water-filled glass vessel is of the same colour as the one that exits the vessel. Water does not absorb light in the visible region of the spectrum; if it did, we would see it coloured due to some missing components of white light. Light absorption is the necessary condition of any photochemical reaction. (This statement, as it was formulated by the great chemist, *Robert Bunsen,* is obvious to us, since we know that light is nothing else but electromagnetic radiation, the energy of which covers the need of the chemical transformation.)

Water absorbs light only in the very short wavelength ultraviolet (UV) region. Here water decomposes developing hydrogen and oxygen. The short-wavelength component of sunlight reaching the Earth surface is much too weak to attach any practical hope to. If, however, some coloured substance is dissolved in water, which absorbs light in the visible region, some photochemical reaction might take place. For example, illuminating aqueous solutions of three-valent cerium salts, a coloured liquid, hydrogen formation is observed but with a very low yield.

This is no wonder. The dissociation products, radicals H· and HO· are produced in each other's close neighbourhood so it is easy for them to re-unite, forming water again. Similarly, if illumination brings about ions and electrons, H_2O^+ and e^-, they neutralize each other. Thus, it is not enough to find a process which dissociates water molecules. Care must also be taken to separate the reaction products. The best thing to do is to make them form far from each other.

Far, all right, but compared to what? Probably to the distance they must cover for a new encounter. This is in the range of several nanometers in solids or liquids (1 nm = 10^{-9} m, the diameter of a water molecule is 0.28 nm). In the chapter on the storage of electric energy the double layers at metal/electrolyte interfaces were mentioned. Their thickness is of that order of magnitude. What if a metal plate immersed into an aqueous solution were illuminated? The double layer might separate the charges formed. It might, if they formed. But no, water is completely transparent and metal is highly reflective of visible light. There is no light absorption, which is the necessary condition of any transformation.

The task to separate the charges is somewhat similar to the rectification of alternating current. To make sure charge carriers move only in a given direction this is what must be achieved in both cases. The scheme of an electric rectifier is shown in Fig. 11. The current, produced by the AC generator, changes its direction of flow regularly. A rectifier switched into the circuit permits the current to flow only in one direction.

Since the advent of semiconductor devices practically only semiconducting diodes are used as rectifiers. A semiconducting substance, e.g. a silicon single crystal, is a good electric insulator in pure state. Similarly to carbon, the Si atom has four valences forming strong covalent bonds in a crystal of diamond-like structure (Fig. 12). Electrons being strongly localized in the covalent bonds they cannot be moved by any external electric field.

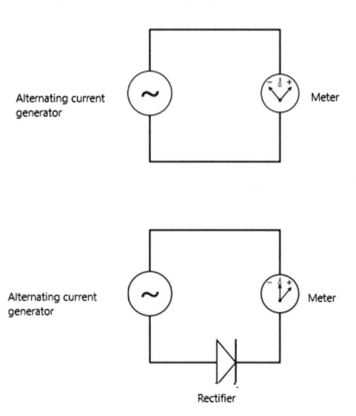

Fig. 11 Scheme of an electric rectifier

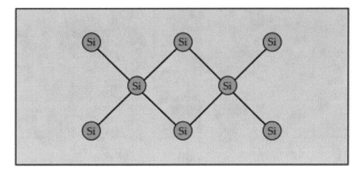

Fig. 12 Simplified structure of a silicon crystal

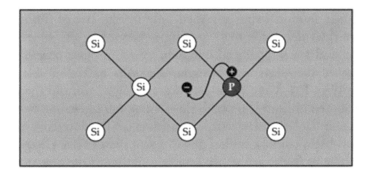

Fig. 13 Silicon crystal with one phosphorous atom

However, silicon ceases to be a high resistance material once a foreign atom, for example a pentavalent phosphorous, is placed into the crystal lattice at a Si site. One sees in Fig. 12. that a P atom has one electron too many for the diamond-type lattice. This excess charge is easily displaced, so the P-containing Si crystal is a conductor with mobile electrons. This kind of semiconductor is called *n*-type, *n* referring to the word 'negative' (Fig. 13).

A similar effect can be achieved by putting a three-valent atom, for example boron, to a silicon site. In that case, one electron is missing from the diamond-type lattice (Fig. 14). A positive charge (i.e. a missing electron, usually denoted a hole) can move freely in the lattice where B was forced to create bonds, with four of its neighbours making the isolating Si crystal an electric conductor. Current is carried here by the positive charge carriers, generally called holes; that is why these types of semiconductors are denoted as *p*-type with reference to the word 'positive'.

Electrons and holes in a crystal resemble positive and negative ions in water. In the absence of any external electric field they move almost at

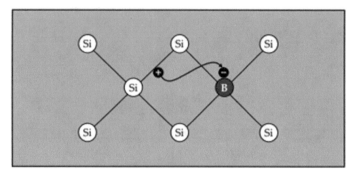

Fig. 14 Silicon crystal with one boron atom

random, their motion being governed only by the law of diffusion: from sites where their density is high they move towards a site where it is low. If a *p*-type and an *n*-type semiconductor come in close contact diffusion takes place at their interface bringing electrons to the *p* side, holes to the *n* side (Fig. 15). Were these particles uncharged diffusion would make them distributed uniformly on both sides. Coulomb force, however, impedes free diffusion. Holes and electrons, enriched at the interface, push their companions back, coming behind due to a repulsion of like charges.

A double layer forms at the *p*/*n* interface similar to the one at the barrier of a metal and an aqueous solution (Fig. 2). Neither electron, nor hole can pass the interface because of the potential threshold brought about by the accumulated charges (Fig. 16).

Such arrangement is called a semiconductor diode, Fig. 16 makes it clear why it behaves as a rectifier. If side *p* is made more negative and side *n* more positive than its spontaneous potential, current cannot flow across the interface. In the opposite case, i.e. with less negative *p* and less positive *n* sides, the current flows freely.

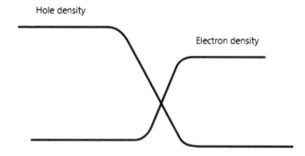

Fig. 15 Charge distribution at the interface of an *n*-type and a *p*-type silicon crystal

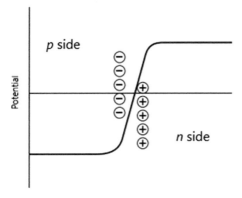

Fig. 16 Potential distribution at a *p/n* interface

A similar method can be made use of in electrochemistry. In that case, a semiconductor is immersed into an aqueous solution. A rectifying layer, similar to that at a *p/n* junction, is formed at the solid/liquid interface. Moreover, the reasons of the two are the same and so are the charges and potentials. An electric double layer is brought about because of the difference of the charge carriers in the contacting phases: electrons *versus* holes at a *p/n* junction, electrons or holes *versus* dissolved ions at a semiconductor/electrolyte interface.

The voltage drop at the semiconductor/liquid interface is given in Fig. 17 together with the scheme of hydrogen production. Let the semiconductor be illuminated with a light which penetrates the liquid but is absorbed by the semiconductor. If the frequency is properly chosen an electron–hole pair is

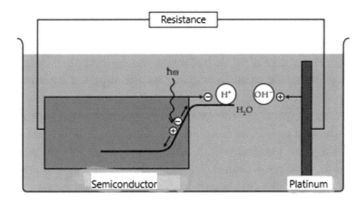

Fig. 17 The double layer at the semiconductor/aqueous solution interface separates the electron–hole pair produced by light within the semiconductor

created in the solid phase,

$$\text{Si crystal} + \hbar\omega \rightarrow \oplus + e^- \tag{5.5}$$

These two would neutralize each other, dissipating the light energy as heat, if no rectifier layer were present. This layer, however, separates the pair of opposite charge carriers. Most often semiconductor and electrolyte are chosen so as to make holes move towards the electrolyte and the electrons into the semiconductor bulk. Electrons travel towards the metal counter electrode along the external circuit. The electrochemical processes close the entire electric circuit. The hole, leaving the semiconductor, reacts with an OH^- ion of water, finally making oxygen develop at the metal electrode,

$$OH^- + \oplus \rightarrow OH \rightarrow \frac{1}{4}O_2 + \frac{1}{2}H_2O \tag{5.6}$$

The electron reacting with a hydrogen ion leaves the metal counter electrode,

$$H^+ + e^- \rightarrow H \rightarrow \frac{1}{2}H_2 \tag{5.7}$$

This is the process wanted: solar energy is transformed to the chemical energy of hydrogen.

The first equipment working along the above principle was constructed by Fujishima and Honda during the energy crisis of the 1970s. Having made the native mineral rutile (titanium dioxide) slightly semi-conductive, this *n*-type semiconductor was immersed into an aqueous solution and switched to a platinum counter electrode. Under sunshine, oxygen developed at the surface of the rutile crystal while hydrogen was formed at the platinum. The cell proved to be stable steadily working on the top of a building of Tokyo University, without eliciting any interest from the industry.

Yield and rate, these are the factors that count; researchers try hard to improve both. The task is far from simple. One has to find a semiconductor that absorbs a good portion of the sunshine, spending it on electron–hole pair formation. The semiconductor surface should be large, its mass high as compared to that of the liquid and it should be a good catalyst for oxygen evolution. The rutile crystal did not comply with all these requirements.

Trying to achieve these aims nanometre sized semiconductor grains were investigated, this dimension being in the order of magnitude of the double layer thickness. An aqueous suspension of such grains behaves as if it consisted of double layers only. Wherever a pair of charge carriers is formed it is gripped and divided by the built-in field of the double layer. The large amount of semiconducting grains behaves as if it were a solution of the semi-conductor. Hence, it is worth illuminating the whole, homogeneous-looking liquid, light finds some solid material everywhere, which can decompose water.

Changing the size of the grains also the light absorption of the substance changes, that is something that one would not expect by knowing only macroscopic objects. In the nanometre range, however, optical spectrum depends on the grain size. Light absorption consists in the increase of electron energy in the solid a photon giving its energy to an electron. Electron energies depend on the nature of the substance, that is why each substance has a different absorption spectrum, i.e. a different colour. If the dimension of a grain is in the order of the optical wavelengths the size also counts. Electrons in solids behave as standing waves, similarly to water in a vessel, the waves of which are reflected by the walls. If the vessel is much larger than the wavelength of the water waves vessel size does not matter. However, if the vessel is as small as or smaller than the wavelength, the size does influence the shape of the waves. Electron wavelengths in solids are in the nanometre range. That is why the electron wavelengths, hence the energies they can absorb, depend on the grain size.

This finding implies the important advantage that as changing the grain sizes light absorption can be fitted to the optical spectrum of the Sun—the semiconductor would absorb a good part of sunlight increasing the efficiency of electron–hole pair production.

Metal counter electrodes are still needed. Parts of the grain surfaces are covered with metal in order to close the electric circuit and maintain continuous water decomposition.

A family of semiconducting substances, called perovskites, are nowadays thoroughly investigated also as far as water decomposition is concerned. These metal-containing, partly organic substances are relatively easy to manu-facture, they have a high energetic yield but their stability and lifetime is not adequate at present. Still, researchers of the field pin high hopes to the family.

Photoelectrochemistry, as this area is called, is a most challenging field which still did not reach the goal set by the needs of energetics.

6   Hydrogen Storage in Solids

One of the advantages of hydrogen as an energy carrier is that it stores energy without any loss. All right. But how do we store hydrogen? That should be simple, safe and cheap. The task is far from being obvious. The density of the substance either as a gas or as a liquid is low, so is its boiling point, hence it is difficult to make it into a liquid. If compressed by heavy steel cylinders at high pressures the economic advantage of the substance's low weight disappears. The aim is to find materials that can contain a reasonable amount of hydrogen at moderate pressures and temperatures and to devise easy and safe ways for the recovery of the element. Cycles of hydrogen binding and release should be developed.

The expression "reasonable amount" refers to two items: appropriate ratios of both volume and weight between hydrogen and the binding material. These two quantities are the axes of the graph in Fig. 18. As set by the US Department of Energy the lowest limits to be attained are 6.5 w/w % and 0.062 hydrogen/l.

The density of hydrogen gas is about 4 w/w % at 450 bar, taking tank weight into account as well. Gas release must be strictly controlled and, although strict rules prescribe the quality of the tank or cylinder, this storage

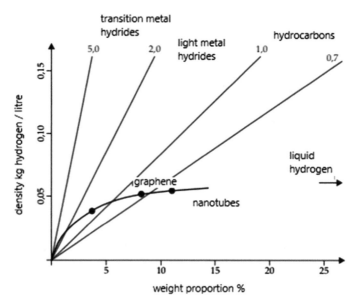

Fig. 18 Weight proportions and volume densities of several hydrogen containing systems. The better the system the nearer its characteristic point to the upper right part of the diagram

is still regarded as dangerous. Liquefaction is safer but the density of liquid hydrogen is as low as 0.068 kg/litre and the boiling point, $-252\,°C$ at 1 bar, is difficult (expensive) to attain. Liquid hydrogen as a fuel is made use of only in space research.

Neither thick walled steel tanks, nor ultralow temperatures are needed if hydrogen is fixed in compounds. A large number of hydrogen-containing compounds are known with a great variety of properties. There is at least one hydrogen compound of almost each element of the periodic table. Some of them are solids with melting points as high as that of salts. Indeed, the structures of these compounds are similar to table salt; while the latter consists of Na^+ and Cl^- ions, sodium hydride, NaH is built of Na^+ and H^- ions. This is an ionic compound with binding electrons residing near the H atoms. Such compounds, similarly to salts, are very stable, e.g. NaH starts to decompose, releasing hydrogen at a temperature of 210 °C. Thus, they are obviously useless for industrial storage. Nevertheless, laboratories might use them as handy hydrogen sources. For example, lithium hydride, LiH releases the gas if water is added,

$$LiH + H_2O \rightarrow LiOH + H_2 \tag{6.1}$$

However, no storage cycle can be performed because Li is not easy to recover.

The compound Li-alanate, $LiAlH_4$ seems to be a promising candidate for storage in view of its high (i.e. more than 10 weight percent) hydrogen content. Increasing and decreasing the pressure periodically over the compound hydrogen can be bound and released. Unfortunately, the reactions are slow and reversibility is poor; now the beneficial effects of some additives to the solid are investigated.

More recently, the hydrogen compounds of boron and nitrogen are regarded as hopeful candidates and so are carbon nanotubes and graphene. These different modifications of carbon seem to have a number of advantages: they are light, stable and bind hydrogen readily and in a reversible way.

As it is well known, a great number of compounds consist of carbon and hydrogen only. The reason of this is the ability of the C atoms to form long chains and the lavish multitude of compounds come from the rich variability of the chains of different lengths, cycles, single and multiple bonds. There is no need to stress the importance of hydrocarbons in a society living on oil and natural gas and producing huge amounts of plastic. As to hydrogen storage, the strength of the carbon-hydrogen bonds is a serious disadvantage. The bonding pair of electrons resides with similar probabilities near the C and the H atom, just like in the case of H_2. This is a stable covalent bond that is almost as difficult to disrupt as the O–H bond, thus it is more challenging to get hydrogen from a hydrocarbon than from water.

There are several solid hydrogen compounds in which the bonds cannot be regarded either ionic or covalent. One is somewhat hesitant to call them compounds since it is well known, *pace* Proust, that the most important feature of a compound is its constant and invariable composition, according to the "the law of definite proportions". Compounds that obey this law are called daltonides, showing little gratitude to Proust. Still, there are compounds the composition of which might change within some limits. They are called berthollides; a reader of the section "Unit of mass" will know the reason why.

Hydrogen and several metals form substances of variable compositions, they can neither be called compounds, nor mixtures. The best studied of them is the palladium/hydrogen system. Metallic palladium can fix hydrogen in an amount the volume of which is almost ten thousand times larger than of the gas. This dazzling figure comes from the low density of hydrogen, meaning only that a phase of the composition Pd_4H_3 can form. This formation goes along with the abrupt change of certain properties of palladium, indicating that it is more of a compound than sheer mixture. Metallic conductor palladium turns into a semiconductor if the hydrogen content reaches or surpasses the composition $PdH_{0.5}$. However, the metal's mechanical properties, e.g. ductility, do not change too much, which is a characteristic of mixtures.

The formation of the Pd/H phase starts with the hydrogen being bound to the surface of the metal and it intrudes the lattice only if the gas pressure is sufficiently high. Then H moves easily and diffuses fast. No other absorbed gas, including helium, move as readily as that, this being the peculiarity of H_2 and D_2. These two are the only gases that can permeate solid palladium, so the industry has much use of the metal for hydrogen purification. The process, although taking place already at room temperature, is, for practical reasons, usually performed around 300 °C.

As far as hydrogen storage is concerned, palladium is of little use, being both expensive and heavy. The high hydrogen content cannot compensate the high weight of the hydrogen + metal system. One has to find metals, lighter and cheaper ones. For example, magnesium hydride, MgH_2 is reasonably light, with the ability to bind or release hydrogen, controlled by gas pressure, in the 200–300 °C interval. Its storage capacity is about 0.055 kg/l being beyond the density of liquid hydrogen. The rate of gas exchange increases if the solid substance is applied as nanometre sized grains.

Several alloys or rather intermetallic compounds are much promising. They might be called compounds because their compositions are strictly defined. Substances made of iron and titanium, FeTi, or of magnesium and nickel, Mg_2Ni, and of lanthanum and nickel, $LaNi_5$ absorb and desorb hydrogen near room temperature with varying external pressure. The density of the stored hydrogen is beyond that of liquid hydrogen.

Resourceful chemists, physicists and engineers, eye catching observations, practical implementations—the work shows good progress. Still, this multitude of results show that the ultimate solution is still not at hand.

Further Reading

A. Godule-Jopek (ed.), *Hydrogen Production: Electrolysis* (Wiley-VCH, London, 2015)

G.F. Naderer, I. Dincer, C. Zamfirescu, *Hydrogen Production from Nuclear Energy* (Springer, London, 2013)

G.A. Olah, A. Goeppert, G.K.S. Prakash, *Beyond Oil and Gas: The Methanol Economy* (Wiley-VCH, London, 2006)

M.H.B. Stiddard, *The Elementary Language of Solid State Physics* (Academic Press, London, 1975)

A. Züttel, A. Borgschulte, L. Schlapbach (eds.), *Hydrogen as a Future Energy Carrier* (Wiley-VCH, Weinheim, 2008)

Conclusions

Having addressed the reader on the first page of this book I promised not to popularize some sort of a textbook, rather to show a way of thinking on questions where the answers are expected from both physical and chemical understanding. I tried to describe the basic ideas and notions of physical chemistry through their history, being of the opinion that it might be easier to recognize the power of discoveries that way. The results of the full-fledged branch of science, however, were discussed much more according to notions and problems detailing some relevant findings only when it was necessary for argumentation, understanding or further applications. That was the reason why the ancient idea of phlogiston and present-day electron theory of oxidation were described in close neighbourhood, or some theorems of reaction kinetics were set out in the chapter on the chemistry of the outer space. Obviously, I was unable to stick to this method of associations all through the book; it is most difficult to be consistent even in inconsistency.

Now, as a conclusion, let us have an overview, according to the chapters of traditional textbooks. The basic notions of *general chemistry* are given in the introductory, historical chapters. The laws of *macroscopic (phenomenological) thermodynamics* are described in the chapters on gases. As a continuation, the explanations of pressure, temperature and specific heat represent the elements of the *kinetic theory of gases.* Going some step further, the basic notions of *quantum mechanics* were formulated and, extending the previous ideas to any substance, the idea of *statistical physics* was introduced. At that point, the basics of *quantum chemistry* (quantum mechanics applied to molecules) were

R. Schiller, *A Non-Traditional Guide to Physical Chemistry,* https://doi.org/10.1007/978-3-031-07488-2_6

introduced. The laws of *chemical thermodynamics* and *chemical kinetics* were discussed in connection with the synthesis and manufacturing of ammonia. Nevertheless, chemical kinetics and related quantum mechanical effects were referred to also in connection with the chemical actions of light and ionizing radiation, i.e. with *photo- and radiation chemistry*. Inconsistency has remained the Leitmotif of the book: the Second Law of thermodynamics appeared again when the problems of hydrogen economy were treated and black body radiation was first mentioned in the chapter on outer-space chemistry.

I fear, there are more of chemical formulae and mathematical expressions in the text than is compatible with the readers's patience. The concluding chapter on hydrogen economy is certainly more detailed and has more of an engineering flavour than the previous ones; perhaps it can be forgiven, because of the present-day importance of the problem. If the reader finds it boring, they might simply skip it. The author must not request each reader to work through the whole book with the same steadfast diligence.

My aim was more ambitious than that. I wanted to show the intellectual technique and skill of a branch of science. To be more honest, I wanted to tell you why I like physical chemistry. Perhaps some readers will feel similarly.

Printed in the United States
by Baker & Taylor Publisher Services